STUDIES IN PROBABILITY AND STATISTICS

Papers in Honour of EDWIN J. G. PITMAN

STUDIES IN PROBABILITY AND STATISTICS

Papers in Honour of EDWIN J. G. PITMAN

Edited by E. J. WILLIAMS
University of Melbourne

NORTH–HOLLAND PUBLISHING COMPANY
AMSTERDAM · OXFORD · NEW YORK

North–Holland ISBN: 0 7204 0434 7

Published by:
North–Holland Publishing Company — Amsterdam
North–Holland Publishing Company, Ltd. — Oxford

Sole distributors for the U.S.A. and Canada:

American Elsevier Publishing Company, Inc.
52 Vanderbilt Avenue
New York, N.Y. 10017

Printed in Israel

CONTENTS

PREFACE

Edwin Pitman has made outstanding contributions to the theory of statistics over a period of more than thirty-five years, and has won the respect and affection of a generation of statisticians. With the approach of his seventy-fifth birthday, in 1972, a group of his colleagues, former students and other associates decided to mark the occasion by assembling a collection of statistical papers for publication in his honour. The present volume is the result.

Pitman's contributions to statistics are characterized by depth of insight, rigour of treatment, clarity of expression and elegance of style. We hope that we have been able to mirror some of these qualities in the papers presented here. This publication is a means of expressing our gratitude to Pitman, who has been, and remains, an inspiring teacher and friend.

E. J. Williams
University of Melbourne

E. J. G. Pitman — A Tribute

E. J. WILLIAMS
University of Melbourne

Edwin Pitman was appointed in 1926 to the Chair of Mathematics of the University of Tasmania, a position he occupied until his retirement at the end of 1962. It was during these years, located in a relatively small and isolated University, that he made his major contributions to the theory of statistics.

Pitman first became interested in applications of statistics through being consulted about the analysis of data by experimentalists of the Tasmanian Department of Agriculture. At that time Fisher's ideas and methods were not widely known outside Britain. Pitman set about mastering Fisher's published works, including "Statistical Methods for Research Workers". As a result, he was able, not only to answer the practical questions he had been presented with, but also to understand the theory underlying the statistical methods. What is remarkable is that, although a mathematician with no experience of experimentation, he grasped the essence of the statistical theory and methods that Fisher had expounded.

His first published papers, which were indicative of the way in which he was able to get to the heart of a problem, were about some aspects of statistical inference which were at that time obscure. In his 1936 paper he gave the essentials of the proof that distributions admitting a sufficient statistic are of exponential type.

Two of his 1939 papers, on inference about location and scale parameters, are perhaps his major contributions to statistical inference. They were written to provide a mathematical framework for the results given in Fisher's 1934 paper, "Two new properties of mathematical likelihood" (Proc. Roy. Soc. London A, **144,** 285–307). In the process they revealed the limitations of simultaneous inferences about location and scale parameters, a result that is apparently still not widely understood. These papers were not readily accepted for publication, because they were at first regarded as too mathematical for British statistical journals. They established the concept of invariance in statistical inference; although the concept has been generalized considerably since then, the results given in these papers are the essential ones with practical relevance.

Pitman early became interested in what inferences could be drawn from samples when no assumptions are made about the form of the population, and this interest led to the three papers on distribution-free tests; they confirmed the view of practical workers that departures from normality did not seriously invalidate the usual tests of significance for means.

To appreciate the nature of Pitman's achievements up to 1940, it must be realized that he was working in virtual isolation; journals then, as now, took weeks to reach Australia from Britain and the United States; there was no airmail, so correspondence was slow, a reply to a letter usually taking at least three months. Pitman did give a course in Theory of Statistics for students taking the Commerce degree; understandably perhaps it attracted only two or three students a year. There was no group of colleagues with whom he could discuss statistical matters; not until his visit to the United States in 1948 did he meet other statisticians.

One effect of this isolation was the risk of being forestalled in publication by others located closer to a centre of research. Pitman's 1940 paper, "A note on normal correlation", was submitted to Biometrika at the same time as a paper on the same topic by W. A. Morgan, who was at University College whence Biometrika was published; fortunately, as the two papers adopted different approaches to the solution, the Editor decided to publish them both.

In 1948–1949 Pitman was invited to visit Columbia University, the University of North Carolina and Princeton University to give lectures on non-parametric inference. The notes of his North Carolina lectures, although unpublished, have been widely circulated in mimeographed form, have often been quoted, and have greatly influenced work in this field.

Pitman's contributions to probability theory and statistics cover a wide range. In his 1937 paper he proved, amongst other things, the special but useful result that the sum of a number of independent gamma variates is distributed independently of any statistic which is homogeneous of degree zero. In his third 1939 paper he proved, as an application of his results, the unbiasedness of Bartlett's test for homogeneity of variances. The concept of asymptotic relative efficiency was introduced in his Columbia lecture notes.

His later work has been concentrated on topics in probability theory. The most important papers in this field are those on the behaviour of characteristic functions in the neighbourhood of the origin; this behaviour determines the nature of limiting distributions of sums of random variables. One result easily derived by this means is what Pitman calls the "eccentric limit theorem"; this states that, if X has a symmetric distribution with continuous and non-vanishing density at zero, then the harmonic mean of independent variates distributed as X has a limiting Cauchy distribution.

Pitman has always been modest about his work, leaving the results to speak for themselves. In his first paper on distribution-free inference he describes his approach as "a bit of a novelty". He has seldom engaged in controversy, believing that it is more profitable to find and publish the facts, and that any conclusions demonstrably false will soon be forgotten. He did, however, express his candid opinions of some aspects of Fisher's work in an article-length review (1957) of Fisher's "Statistical Methods and Scientific Inference". Later, in a discussion of Fiducial Probability at the 1963 Ottawa Session of the International Statistical Institute, he expressed his views trenchantly but without rancour.

Edwin Pitman has been honoured by being elected a Fellow of the Australian Academy of Sciences, a Fellow of the Institute of Mathematical Statistics, a Member of the International Statistical Institute, an Honorary Fellow of the Royal Statistical Society, an Honorary Life Member of the Statistical Society of Australia and an Honorary Life Member of the Australian Mathematical Society.

Since his retirement, he has travelled widely and has been invited to visiting professorships, at the University of California at Berkeley and Johns Hopkins University (1963), the University of Adelaide (1965), the University of Melbourne (1966 and 1967), the University of Chicago (1969), and the University of Dundee (1973). He retains an undiminished keenness of mind and interest in statistical theory.

BIBLIOGRAPHY OF E. J. G. PITMAN

1936 Sufficient statistics and intrinsic accuracy. *Proc. Camb. Phil. Soc.* **32**, 567–79.

1937a The "closest" estimates of statistical parameters. *Proc. Camb. Phil. Soc.* **33**, 212–22.

1937b Significance tests which may be applied to samples from any populations. I. II. The correlation coefficient test. *J. R. Statist. Soc. Suppl.* **4**, 119–30; 225–32.

1938 Significance tests which may be applied to samples from any populations. III. The analysis of variance test. *Biometrika* **29**, 322–35.

1939a The estimation of the location and scale parameters of a continous population of any given form. *Biometrika* **30**, 391–421.

1939b A note on normal correlation. *Biometrika* **31**, 9–12.

1939c Tests of hypotheses concerning location and scale parameters. *Biometrika* **31**, 200–15.

1949a *Lecture Notes on Non-parametric Statistical Inference*. Columbia University.

1949b (with Herbert Robbins) Application of the method of mixtures to quadratic forms in normal variates. *Ann. Math. Statist.* **20,** 552–60.

1956 On the derivatives of a characteristic function at the origin. *Ann. Math. Statist.* **27,** 1156–60.

1957 Statistics and science. *J. Am. Statist. Ass.* **52,** 322–30.

1960 Some theorems on characteristic functions of probability distributions. *Proc. Fourth Berkeley Symposium on Math. Statist. and Prob.* **II,** 393–402.

1965 Some remarks on statistical inference. *Bernoulli* (1723) *Bayes* (1763) *Laplace* (1813), 209–16. Berlin: Springer.

1967 (with E. J. Williams) Cauchy-distributed functions of Cauchy variates. *Ann. Math. Statist.* **38,** 916–8.

1968 On the behaviour of the characteristic function of a probability distribution in the neighbourhood of the origin. *J. Aust. Math. Soc.* **8,** 423–43.

1972 Simple proofs of Steck's determinantal expressions for probabilities in the Kolmogorov and Smirnov tests. *Bull. Aust. Math. Soc.* **7,** 227–32.

Multivariate Binomial Distributions

H. O. LANCASTER
University of Sydney

1. Introduction

Francis Galton, W. F. R. Weldon, Karl Pearson and many others have sought to explain stochastic dependence by the notion that the random variables in correlation are sums, which have elementary random elements in common. There are genetic examples which render this notion very attractive. However, although the mathematical theory is well known for the normal distribution, there seems to be no general theory of joint binomial distributions. Indeed, it is not well known that simple models can give rise to different types of joint distribution. All the necessary results can be derived from the observation that the orthogonal polynomials on the binomial distribution named after Kravčuk (Krawtchouk, 1929) can be represented as sums of products, r at a time, of the elementary summands where r is the degree of polynomial.

In all the cases considered below, where the marginal variables are generated by the addition of elementary summands, the probability function can be expanded in a diagonal polynomial expansion generalizing the Mehler expansion of the normal distribution. It is of interest that as the indices of the marginal variables tend to infinity, the Kravčuk polynomials pass over into the Hermite and the canonical correlations tend to the powers of the first of them. In the joint distribution of several marginal binomial variables, terms appear, which have no analogue in the generalized Mehler expansion. These vanish in the limiting distributions because the norming factors introduce powers of $N^{-\frac{1}{2}}$ into the coefficients.

2. The Kravčuk polynomials

The polynomials on the binomial distribution with index, N, and parameter, p, may be defined by the generating function,

$$1 + \sum t^r P_r(X^*) \equiv K(X^*, t; N) = (1+qt)^{X^*}(1-pt)^{N-X^*}, \qquad p+q=1. \tag{2.1}$$

However, the binomial variable, X^*, can be represented as a sum,

$$X^* = X_1^* + X_2^* + \cdots + X_N^*, \tag{2.2}$$

of mutually independent summands, the *elementary binomial variables*, each of which has an index of unity and a parameter p. $K(X^*, t; N)$ of (2.1) can thus be represented as a product,

$$K(X^*, t; N) = (1+qt)^{\Sigma X_k^*}(1-pt)^{N-\Sigma X_k^*} = \prod_1^N \left[(1+qt)^{X_k^*}(1-pt)^{1-X_k^*}\right] = \prod_1^N \left[1 + t(X_k^* - p)\right]. \tag{2.3}$$

A comparison of (2.1) and (2.3) shows that

$$P_r(X^*) = \sum_{k_i \neq k_j} (X_{k_1}^* - p)(X_{k_2}^* - p)\cdots(X_{kr}^* - p). \tag{2.4}$$

We now write

$$X_k = (X_k^* - p)p^{-\frac{1}{2}}q^{-\frac{1}{2}}, \tag{2.5}$$

and

$$X = (X^* - Np)(Npq)^{-\frac{1}{2}}. \tag{2.6}$$

Each X_k is now standardized to have zero expectation and unit variance. Similarly X is standardized. We write the rth orthonormal polynomial in X as $x^{(r)}$,

$$x^{(r)} = \binom{N}{r}^{-\frac{1}{2}} \Sigma X_{k_1} X_{k_2} \cdots X_{k_r}, \tag{2.7}$$

where $k_i \neq k_j$ if $i \neq j$. It has already been shown that the sum is a polynomial of precise degree r, by means of the definition of $P_r(X^*)$. It is easily verified that $x^{(r)}$ is also correctly standardized by using the mutual independence and the orthonormal properties of the X_k, and so when the sum is squared, the cross products all have zero expectation.

3. Elementary binomial and multinomial summands

We define the *elementary binomial variables* or *summands* by

$$\mathscr{P}(X_k^* = 1) = p_X, \qquad \mathscr{P}(Y_k^* = 1) = p_Y, \qquad \cdots \tag{3.1}$$

where $p_X, p_Y, p_Z, \cdots$ remain fixed for every k. In the models to be given below the elementary summands are supposed to have joint distributions which are characterized by the ordinary coefficients of correlation,

$$\rho_{XY} = \mathscr{E}(X_k Y_k), \qquad \rho_{XZ} = \mathscr{E}(X_k Z_k), \cdots \tag{3.2}$$

and the generalized coefficients of correlation of Lancaster (1960),

$$\rho_{XYZ} = \mathscr{E}(X_k Y_k Z_k), \qquad \rho_{XYZW} = \mathscr{E}(X_k Y_k Z_k W_k), \cdots. \tag{3.3}$$

Since the distributions are defined on a finite set of points, the joint distribution of $X_k, Y_k, Z_k, \cdots$ is completely determined by $p_X, p_Y, p_Z, \cdots$ together with (3.2) and (3.3). We call the collection of variables X_k, Y_k, Z_k a *joint summand*.

In all models to be introduced $p_X, p_Y, p_Z, \cdots$ are fixed quantities independent of k and do not take values of 0 or 1. Where the coefficients of correlation of (3.2) and (3.3) vary with k, this is indicated by including an additional subscript k, as in $\rho_{XY,k}$.

We refer to the distribution of the *joint summand* as the *elementary multivariate binomial distribution*, of which a special case is the *elementary multinomial distribution*.

4. The bivariate binomial distributions

We consider now four models by which joint distributions of binomial marginal variables can be generated. Only in the first of these models is it necessary to assign equal probabilities to the elementary variables, p_X and p_Y. In any series of trials, p_X and p_Y are independent of the ordinal number of the trial.

Model 1. *Random elements in common.* We take as elementary variables or summands $U_1, U_2, \cdots$, the elementary binomials in standard form, so that $\mathscr{P}(U_k = -p^{\frac{1}{2}}q^{-\frac{1}{2}}) = q$ and $\mathscr{P}(U_k = p^{-\frac{1}{2}}q^{\frac{1}{2}}) = p$, $p + q = 1$. We write

$$(n_1 + n_2)^{\frac{1}{2}} X = U_1 + U_2 + \cdots + U_{n_1+n_2},$$
$$(n_2 + n_3)^{\frac{1}{2}} Y = U_{n_1+1} + U_{n_1+2} + \cdots + U_{n_1+n_2+n_3}. \tag{4.1}$$

The n_2 elements, $U_{n_1+1}, U_{n_1+2}, \cdots, U_{n_1+n_2}$, are held in common. Then according to the analysis of §2, the orthonormal polynomials on the distribution of X can be written as

$$x^{(r)} = \binom{n_1 + n_2}{r}^{-\frac{1}{2}} \Sigma\, U_{i_1} U_{i_2} \cdots U_{i_r}, \qquad i_1 > i_2 > \cdots > i_r, \tag{4.2}$$

where summation is over all such r-combinations of the integers, $1, 2, \cdots, n_1 + n_2$. Similarly,

$$y^{(s)} = \binom{n_2 + n_3}{s}^{-\frac{1}{2}} \Sigma U_{j_1} U_{j_2} \cdots U_{j_s}, \tag{4.3}$$

where summation is over all s-combinations of the integers $n_1 + 1, n_1 + 2, \cdots, n_1 + n_2 + n_3$. The expectations $\mathscr{E}(x^{(r)} y^{(s)})$ can be readily evaluated, for in the product of the sums, the cross products have zero expectation unless $r = s$ and $i_k = j_k$ and there are $\binom{n_2}{r}$ such combinations, so that

$$\mathscr{E}(x^{(r)} y^{(s)}) = \binom{n_1 + n_2}{r}^{-\frac{1}{2}} \binom{n_2 + n_3}{s}^{-\frac{1}{2}} \delta_{rs} \binom{n_2}{r} = (n_2)_r / [(n_1 + n_2)_r (n_2 + n_3)_r]^{\frac{1}{2}}, \tag{4.4}$$

where $(n_i)_r = n_i(n_i - 1) \cdots (n_i - r + 1)$, a result which can also be obtained by the use of generating functions after the manner of Eagleson (1964). In particular,

$$\rho \equiv \mathscr{C}(X, Y) = n_2 / [(n_1 + n_2)(n_2 + n_3)]^{\frac{1}{2}}, \tag{4.5}$$

and it is evident that ρ measures the number or proportion of elementary summands held in common. There are precisely n_2 canonical correlations.

The joint moment generating function of the unstandardized variables, X^* and Y^*, is of the form,

$$(pe^t + q)^{n_1} (pe^{t+u} + q)^{n_2} (pe^u + q)^{n_3}, \tag{4.6}$$

and the joint probability generating function is of the form

$$(pt + q)^{n_1} (ptu + q)^{n_2} (pu + q)^{n_3}. \tag{4.7}$$

The joint probability function can be evaluated as

$$\mathscr{P}(X^* = r, Y^* = s) = \sum_{k=0}^{\min(r,s)} \mathscr{P}(W_1^* = r - k) \mathscr{P}(W_3^* = s - k) \mathscr{P}(W_2 = k), \tag{4.8}$$

where $X^*, Y^*, W_1^*, W_2^*, W_3^*$ are binomial variables in unstandardized form with indices, $n_1 + n_2, n_2 + n_3, n_1, n_2$ and n_3 respectively.

Model 2. Correlated elementary summands. The two marginal variables, X and Y, are standardized sums of elementary variables X_k and Y_k respectively, $k = 1, 2, \cdots, N$, and similarly for Y_k; $\mathscr{C}(X_k, Y_k) = \rho$. The pairs $(X_1, Y_1), (X_2, Y_2), \cdots, (X_N, Y_N)$ are mutually independent. $X = N^{-\frac{1}{2}} \Sigma_1^N X_k$, $Y = N^{-\frac{1}{2}} \Sigma_1^N Y_k$. The probability generating function of X and Y was given by Aitken & Gonin (1935) in the form,

$$(f_{00} + f_{10} t + f_{01} u + ut f_{11})^N, \quad \Sigma f_{ij} = 1. \tag{4.9}$$

The correlation of the elementary summands is given by

$$\rho = (f_{11} - f_{1.}f_{.1})(f_{1.}f_{0.}f_{.1}f_{.0})^{-\frac{1}{2}}, \tag{4.10}$$

where $f_{11} - f_{1.}f_{.1} = \delta$ in their notation.

We give an alternative derivation of the joint distribution of X and Y.

$$\mathscr{E}(x^{(r)}y^{(s)}) = \binom{N}{r}^{-1} \delta_{rs}\mathscr{E}\ \Sigma\, X_{i_1}Y_{i_1}X_{i_2}Y_{i_2}\cdots X_{i_r}Y_{i_r} \tag{4.11}$$

since, in the expansion of the product of the sums of the form (2.7) in the X's and the Y's, any term will vanish if $r \neq s$ or if the indices of the X's and Y's are not identical. ρ^r will thus appear $\binom{N}{r}$ times and so

$$\mathscr{E}x^{(r)}y^{(s)} = \delta_{rs}\rho^r. \tag{4.12}$$

The joint probability function of X and Y is determined by its marginal distributions and the coefficients of correlation and can be written

$$f(x,y) = b(x;p_X,N)b(y;p_Y,N)(1 + \sum_1^N \rho^r x^{(r)}y^{(r)}), \tag{4.13}$$

which is of the same form as the Mehler expansion of the joint normal distribution.

Model 3. *The multinomial model.* In this model, which we have termed the multinomial model rather than the multinomial distribution, the joint distribution in two variables is generated by

$$[p_X t_1 + p_Y t_2 + (1 - p_X - p_Y)]^N. \tag{4.14}$$

The coefficient of correlation of the elementary summands, X_k and Y_k, is given by

$$\rho = -(p_X p_Y / [(1-p_X)(1-p_Y)])^{\frac{1}{2}}. \tag{4.15}$$

This model can be reconciled with Model 2 by setting

$$f_{00} = (1 - p_X - p_Y),\ f_{10} = p_X,\ f_{01} = p_Y \text{ and } f_{11} = 0.$$

The joint probability is then given by (4.13).

Model 4. *The generalized model.* In this model, it is assumed that the joint summands are mutually independent. For the X's the sampling model is Bernoullian and similarly for the Y's. It is not assumed that $p_X = p_Y$ except when the correlation at some trial is unity. Let the correlations be $r_1, r_2, \cdots, r_N$. Then the canonical correlations are given by

$$R_1 = \binom{N}{1}^{-1} \Sigma\, r_i$$

$$R_2 = \binom{N}{2}^{-1} \sum_{i<j} r_i r_j$$

$$R_3 = \binom{N}{3}^{-1} \sum_{i<j<k} r_i r_j r_k \tag{4.16}$$

. . . .

By substituting n_2 values of unity into (4.16), we obtain (4.5) with $N = n_2 + n_1 = n_2 + n_3$. By setting $r_1 = r_2 = \cdots = r_N$, we obtain (4.12) and by the specialization

$$r_1 = -(p_X p_Y / [(1-p_X)(1-p_Y)])^{\frac{1}{2}},$$

(4.13) is obtained.

Relation between models. A reconciliation of Models 1, 2 and 4 is more readily obtained if the joint distributions of the elementary summands are defined so that the marginal probabilities remain fixed although the correlations among the elements of an elementary summand are allowed to vary between elementary summands and simple sums are replaced by expressions of the form, $\Sigma \varepsilon_{Xk} X_k^*$, where $\varepsilon_{Xk} = 0$ or 1. ε_{Xk} and ε_{Yk} are not necessarily equal. Thus in Model 1, we should have $\rho_{XY,k} = \rho_{XY} = 1$ and $\varepsilon_{Xk} = 1$ for $k = 1, 2, \cdots, n_1 + n_2$, $\varepsilon_{Xk} = 0$ otherwise; $\varepsilon_{Yk} = 1$ for $k = n_1 + 1, n_1 + 2, \cdots, n_1 + n_2 + n_3$. $\varepsilon_{Yk} = 0$ otherwise.

5. Multivariate binomial distributions

Generalizations of the models given in §4 to several dimensions are possible. With the notation of (3.2) and (3.3), the joint distribution of X, Y and Z may be determined. The expectations of $x^{(r)}y^{(s)}$, $x^{(r)}z^{(t)}$ and $y^{(s)}z^{(t)}$ for $r, s, t > 0$ will remain the same in all models as for the bivariate cases. The expectation of $x^{(r)}y^{(s)}z^{(t)}$ will be given by

$$\rho_{rst} = \mathscr{E}\left[\binom{N_X}{r}^{-\frac{1}{2}}\binom{N_Y}{s}^{-\frac{1}{2}}\binom{N_Z}{t}^{-\frac{1}{2}} \Sigma X_{i_1}X_{i_2}\cdots X_{i_r}\Sigma Y_{j_1}Y_{j_2}\cdots Y_{j_s}\Sigma Z_{k_1}Z_{k_2}\cdots Z_{k_t}\right]. \tag{5.1}$$

The evaluation of this sum is a combinatorial problem. There will be expectations of products of the forms, $X_k Y_k$, $X_{k'} Y_{k'} Z_{k'}$ to be considered.

Examples. (*i*) Suppose the joint distribution of the elementary summands is such that $\rho_{XY,k}$, $\rho_{XZ,k}$ and $\rho_{YZ,k}$ are zeros for all k but $\rho_{XYZ,k} = \rho$; and in Model 5 $\varepsilon_{Xk} = \varepsilon_{Yk} = \varepsilon_{Zk} = 1$, $k = 1, 2, \cdots, N$. Then the two-variable correlations in the joint distribution of X, Y and Z are all zeros. However,

$$\rho_{rst} = \binom{N}{r}^{-\frac{3}{2}}\binom{N}{r}\rho^r = \binom{N}{r}^{-\frac{1}{2}}\rho^r, \tag{5.2}$$

if $r = s = t$ and $\rho_{rst} = 0$ otherwise.

(*ii*) Suppose $\rho_{XY\cdot k} = \rho_3$, $\rho_{XZ\cdot k} = \rho_2$ and $\rho_{YZ,k} = \rho_1$, $\rho_{XYZ,k} = \rho$ for all k and that X, Y and Z are simple sums as in the pfreceding example. Then the coefficient of $\rho_1^i\rho_2^j\rho_3^k\rho^l$ is obtained by summation over all distinct sets of indices such that $j + k + l = r$, $i + k + l = s$, $i + j + l = t$, which can be chosen in $N!/(i!j!k!l!)$ ways, and so from (5.1)

$$\rho_{rst} = \binom{N}{r}^{-\frac{1}{2}}\binom{N}{s}^{-\frac{1}{2}}\binom{N}{t}^{-\frac{1}{2}} N! \sum_{i,j,k} \rho_1^i\rho_2^j\rho_3^k\rho^l/(i!j!k!l!), \tag{5.3}$$

subject to $j + k + l = r$, $i + k + l = s$, $i + j + l = t$. It may be remarked that in (5.2) and in (5.3), where $l > 0$, the coefficient of the term containing ρ^l is $O(N^{-\frac{1}{2}l})$ as N becomes large and such terms play no part in the asymptotic distribution. The evaluation of the coefficients when the $\rho_{XY,k}, \cdots, \rho_{XYZ\cdot k}$ are independent of k is simple; when these coefficients depend on k, some combinatorial problems arise.

6. The null distribution in paired comparisons

A sixth model yields the joint distribution of the number of preferences accorded to objects $A_1, A_2, \cdots, A_n$ in paired comparisons under the null hypothesis. The elementary summands take the values ± 1, and the pairs constituting a paired comparison are mutually independent; each p is $\frac{1}{2}$. Each object appears in precisely $(n-1)$ of the $\binom{n}{2}$ pairs and in precisely one pair with any other given object. The convention is adopted that if the first object does not appear in the pair, the corresponding elementary summand, X_k^* say, is zero. The paired comparison table is usually written with zeros and units (as by David, 1963, on his page 21). The marginal variables are the sums of the entries, $X^*, Y^*, \cdots, Z^*$, in the first, second, $\cdots$, nth rows.

The joint distribution of X^* and Y^* can be written out with the aid of Model 4 above. There is a single correlation of -1 between the elementary summands, X_1^* and Y_1^* say. The joint frequency is given by

$$\begin{aligned} f(x^*, y^*) &= b(x^*; \tfrac{1}{2}, n-1) b(y^*; \tfrac{1}{2}, n-1)[1 - (n-1)^{-1} x^{(1)} y^{(1)}] \\ &= b(x^*) b(y^*)[1 - (n-1)^{-1} x^{(1)} y^{(1)}]. \end{aligned} \tag{6.1}$$

With three variables,

$$\begin{aligned} f(x^*, y^*, z^*) = b(x^*)b(y^*)b(z^*)\Big[1 - (n-1)^{-1}(x^{(1)}y^{(1)} + x^{(1)}z^{(1)} + y^{(1)}z^{(1)}) \\ + \binom{n-1}{2}^{-\frac{1}{2}} (n-1)^{-1}(x^{(2)}y^{(1)}z^{(1)} + x^{(1)}y^{(2)}z^{(1)} + x^{(1)}y^{(1)}z^{(2)}) - \binom{n-1}{2}^{-\frac{3}{2}} x^{(2)}y^{(2)}z^{(2)} \Big]. \end{aligned} \tag{6.2}$$

The results can be generalized with the aid of a correlation generating function,

$$G(\{u_k\}) = \prod_{i<j} (1 - u_i u_j) \equiv \Sigma \rho_{rs\cdots t} \left[\binom{n-1}{r} \binom{n-1}{s} \cdots \binom{n-1}{t} \right]^{\frac{1}{2}} u_1^r u_2^s \cdots u_n^t \tag{6.3}$$

and

$$f(x, y, \cdots, z) = b(x)b(y) \cdots b(z)(1 + \Sigma \rho_{rs\cdots t} x^{(r)} y^{(s)} z^{(t)}). \tag{6.4}$$

With m repetitions, the correlation generating function becomes

$$G(\{u_k\}, m) = \prod_{i<j} (1 - u_i u_j)^m \equiv \Sigma\, \rho_{rs\cdots t} \left[\binom{nm-m}{r} \binom{nm-m}{s} \cdots \binom{nm-m}{t} \right]^{\frac{1}{2}} u_1^r u_2^s \cdots u_n^t, \tag{6.5}$$

and the expansion (6.4) remains the same in form but the indices of the binomials are now all $(nm - m)$ and the orthogonal polynomials depend on these new parameters.

Ties are readily treated. Each tie reduces the indices of the two binomials by unity and deletes a term in the product form of the generating functions of (6.3) and (6.5).

7. Discussion

The above models cover all types of joint distribution, so far described in the literature, obtained by taking sums of independent summands, as by Aitken & Gonin (1935), Eagleson (1964) and Krishnamoorthy (1951). We may remark that (1.3.2) of Krishnamoorthy (1951) gives a generalization to several dimensions not consistent with ours. The canonical variables of our Model 3 are obtained elegantly by Csáky & Fischer (1960).

Asymptotic distributions can be obtained by allowing $N \to \infty$ and the $X, Y, \cdots$ are asymptotically normal and the Kravčuk polynomials pass over into the Hermite. It is of interest that in so doing all information about the original variables is lost except the mean first order correlations. Similar remarks apply to the Poisson limiting process, when the Kravčuk polynomials pass over into the Poisson-Charlier polynomials.

References

AITKEN, A. C. & GONIN, H. T. (1935). On fourfold sampling with and without replacement. *Proc. Roy. Soc. Edinb.* **55,** 114–25.

CSÁKI, P. & FISCHER, J. (1960). Contributions to the problems of maximal correlation. *Magyar Tud. Akad. Mat. Kutato Int. Kozl.* **5,** 325–37.

DAVID, H. A. (1963). *The Method of Paired Comparisons.* London: Griffin.

EAGLESON, G. K. (1964). Polynomial expansions of bivariate distributions. *Ann. Math. Statist.* **35,** 1208–15.

KRAWTCHOUK, W. (1929). Sur une généralisation des polynômes d'Hermite. *Acad. Sci. Paris* **189,** 620–2.

KRISHNAMOORTHY, A. S. (1951). Multivariate binomial and Poisson distributions. *Sankhyā* **11,** 117–24.

LANCASTER, H. O. (1960). On statistical independence and zero correlation in several dimensions. *J. Austral. Math. Soc.* **1,** 492–6.

Some Limit Theorems for Markov Chains with Applications to Branching Processes

A. G. PAKES
Monash University

1. Introduction

Let $\{X_n\}_{-\infty}^{\infty}$ be a stationary sequence of random variables defined on a probability space $(\Omega, \mathscr{A}, P)$. The sequence is said to satisfy the central limit property (C.L.P.) if there exist constants a_n, b_n such that $(\sum_{m=0}^{n} X_m - a_n)/b_n$ is asymptotically normally distributed. Much effort has been devoted to the problem of finding conditions ensuring the validity of the C.L.P.. One aspect of this work has been to determine measures of dependence between certain subsequences of $\{X_n\}$ such that if these measures become small in a certain sense, the C.L.P. will apply. For example, let $\mathscr{B}_n$ (resp. $\mathscr{F}_n$) be the σ-field generated by $\{X_m\}_{-\infty}^{n}$ (resp. $\{X_m\}_n^{\infty}$) and let

$$\alpha(n) = \sup_{\substack{A \in \mathscr{B}_0 \\ B \in \mathscr{F}_n}} \left| P(AB) - P(A)P(B) \right|.$$

If $\alpha(n) \to 0$ $(n \to \infty)$, the sequence is said to be strongly mixing and it is known that if this convergence is sufficiently rapid and certain moment conditions are satisfied, the C.L.P. holds (Ibragimov & Linnik, 1971, p. 334). Again, if

$$\beta(n) = \sup_{\substack{A \in \mathscr{B}_m \\ B \in \mathscr{F}_{m+n}}} \left| P(AB) - P(A)P(B) \right| / P(A) \to 0 \qquad (n \to \infty),$$

the sequence is said to be uniformly mixing. This is more restrictive than the strong mixing assumption, and slightly weaker moment assumptions suffice for the C.L.P. (Ibragimov & Linnik, 1971, p. 340). See Serfling (1968) for further discussion of measures of dependence.

Suppose now that $\{X_n\}$ is an aperiodic stationary Markov chain with state-space $\mathscr{S}$, transition probabilities $p^{(n)}(\zeta, A) = P(X_{m+n} \in A \mid X_m = \zeta)$, $\zeta \in \mathscr{S}$, A a Borel subset of $\mathscr{S}$ and stationary measure $\pi(A)$ such that

$$\sup_{\zeta, A} \left| p^{(n)}(\zeta, A) - \pi(A) \right| \leqq C\rho^n, \tag{1}$$

where C, ρ are constants, $0 < \rho < 1$. Then, as is shown in Ibragimov & Linnik (1971, p. 367), the Markov chain is uniformly mixing and second order moment assumptions imply the C.L.P. for $\{f(X_n)\}$, $f(\cdot)$ being a measurable function on $\mathscr{S}$. It is further shown that the C.L.P. will hold for any initial distribution.

Nagaev (1957) starts with condition (1) and obtains the C.L.P. by investigating the spectral properties of an operator defined by the kernel $P(\zeta, A, \theta) = E(e^{i\theta f(X_{n+1})}; X_{n+1} \in A \mid X_n = \zeta)$. Finally, we mention that Rosenblatt (1971) obtains the C.L.P. for $\{f(X_n)\}$ under a weaker dependence condition than (1), and for the case of a countable $\mathscr{S}$ we refer to Chung (1967) for a proof of Doeblin's renewal theory approach.

It is the case of countable $\mathscr{S}$ that concerns us in this paper and with this restriction we obtain the C.L.P. using elementary methods based on Vere-Jones' elegant theory of non-negative matrices. This approach is, perhaps, closest to that of Nagaev but we do not make use of (1). Indeed, in the present context, (1) becomes a stringent form of geometric ergodicity and although we do not specifically require this latter condition, it is the case that the chains to which we apply our results are, in fact, geometrically ergodic, but it is not known if they satisfy (1). Much of the analysis is based on real variable methods which forces the use of moment generating functions (m.g.f.'s). However, a new continuity theorem for m.g.f.'s, presented in §2, allows us to avoid the requirement that they be defined in open sets containing the origin, that is, the distributions of interest need have only one exponentially decreasing tail (the lower one).

More specifically, we shall be interested in two particular situtations, namely C_1, the conventional case where $\mathscr{S}$ will form an irreducible class of positive recurrent states and the following situation, C_2. Let $\mathscr{I}$ be a countable proper subset of $\mathscr{S}$ and suppose that $\mathscr{I}$ is irreducible. Furthermore, setting $p_{ij} = P(X_{n+1} = j \mid X_n = i)$ $(i, j \in \mathscr{S})$, we suppose that $\sum_{j \in \mathscr{I}} p_{ij} < 1$ for at least one $i \in \mathscr{I}$ and $p_{ij} = 0$ $(i \notin \mathscr{I},\, j \in \mathscr{I})$. Thus $\mathscr{I}$ is a transient set of states from which escape to the absorbing set $\mathscr{S} \setminus \mathscr{I}$ occurs with positive probability. Let $T_i = \inf\{n \mid X_n \notin \mathscr{I},\, X_0 = i\}$ be the time to absorption, and for a general initial distribution we denote this by T. In both cases we consider sums of the form $S_n = Y_0 + Y_1 + \cdots + Y_n$ $(n = 0, 1, \cdots)$, where Y_m $(m = 1, 2, \cdots)$ is a random variable whose distribution function (D.F.) depends on X_{m-1} and X_m, and the D.F. of Y_0 depends on X_0, and given $X_0, \cdots, X_n$ the sequence $\{Y_0, \cdots, Y_n\}$ is conditionally independent. For C_1 we will present a C.L.P. for $\{Y_n\}$. We mention that a similar approach has been adopted by Matthews (1971) for a multivariate C.L.P. for vector-valued functionals on a Markov chain, but that very stringent moment assumptions are made and the proof appears to be incomplete.

For C_2 we shall obtain in §3 conditional laws of large numbers and central limit theorems for S_n given that $T > n$, or given $T = m + n$ or given $T > m + n$ with $m \to \infty$ and then $n \to \infty$. A C.L.P. for the first-mentioned conditioning event was given by Matthews (1970) for functionals on a finite state Markov chain. The properties of X_n given the first and third conditioning events have been elucidated in the well known work of Seneta & Vere-Jones (1966), and for the second conditioning event see Pakes (1973 b). In §4 our limit theorems will be applied to the subcritical Galton-Watson process with, and without, immigration. For C_2 some restriction on the initial distribution is required for the proofs to be valid. Examples will be given in §3 which show that, at least for branching processes, the restriction is necessary.

Continuous time analogues of the results in §3 are presented in §5. These are preceded by a discussion on the extension to Markov processes of the Seneta-Vere-Jones and Pakes theorems, referred to above. In §6 our limit theorems are applied to Markov branching processes with, and without, immigration. The proofs in these sections depend on the corresponding discrete time theorems and embedding techniques based on Croftian theorems; see Kingman (1963 a).

2. A CONTINUITY THEOREM

Let $\{F_n(x)\}$ $(n = 1, 2, \cdots)$ be a sequence of D.F.'s and let $\phi_n(\theta) = \int_{-\infty}^{\infty} e^{-\theta x} dF_n(x)$ be the corresponding m.g.f.'s. If the $F_n(x)$ are concentrated on $[0, \infty)$ the m.g.f.'s become

Laplace-Stieltjes transforms and the continuity theorem for this case is well known, e.g. Feller (1971). When the D.F's are not so restricted but the $\phi_n(\theta)$ are finite in an interval of the form $(-b, b)$, $0 < b \leqq \infty$ a number of continuity theorems are known, the best version being due to Kozakiewicz (1947). We now use both these theorems to obtain a continuity theorem for the case where the $\phi_n(\theta)$ are only assumed finite in $[0, b)$.

Theorem 1. *If the* $\phi_n(\theta)$ *are defined as above, are finite in* $[0, b)$, *and* $\phi_n(\theta) \to \phi(\theta)$ $(n \to \infty; 0 \leqq \theta \leqq b)$, *where* $\phi(\theta)$ *is an m.g.f. which is finite for* $-b < \theta < b$ *then*, $F_n(x) \to F(x)$ $(n \to \infty)$ *at all continuity points of* $F(x)$, *which is the distribution function such that* $\phi(\theta) = \int_{-\infty}^{\infty} e^{-\theta x} dF(x)$.

Proof. By Helly's first Theorem we may select a subsequence $\{n'\}$ such that $F_{n'}(x) \to F(x)$ at continuity points of the latter, which is a non-decreasing right-continuous function. Letting $G_n(x) = F_n(x)$ $(x \geqq 0)$, $= 0$ $(x < 0)$; $H_n(x) = 1$ $(x \geqq 0)$, $= F_n(x)$ $(x < 0)$ and $U(x) = 1$ $(x \geqq 0)$, $= 0$ $(x < 0)$, we have $F_n(x) = G_n(x) + H_n(x) - U(x)$ and $G_{n'}(x) \to G(x)$, $H_{n'}(x) \to H(x)$, where $G(x)$ and $H(x)$ are possibly improper D.F.'s concentrated in (or on) $[0, \infty)$ and $(-\infty, 0]$, respectively.

Let $\gamma_n(\theta) = \int_{0-}^{\infty} e^{-\theta x} dG_n(x)$ and $\psi_n(\theta) = \int_{-\infty}^{0+} e^{-\theta x} dH_n(x)$ so that $\phi_n(\theta) = \gamma_n(\theta) + \psi_n(\theta) - 1$ and $\psi_n(\theta)$ is an m.g.f. finite for $-\infty < \theta < b$. The Helly-Bray Theorem yields $\gamma_{n'}(\theta) \to \gamma(\theta)$ $= \int_{0-}^{\infty} e^{-\theta x} dG(x)$ $(\theta > 0)$ so that for $0 < \theta < b$, $\psi_{n'}(\theta) \to \psi(\theta) = \phi(\theta) - \gamma(\theta) + 1$. For $\theta < 0$, we have $\psi_{n'}(\theta) \to \int_{-\infty}^{0+} e^{-\theta x} dH(x)$ and we let this define $\psi(\theta)$ for negative θ. Finally set $\psi(0) = 1$, so that $\{\psi_{n'}(\theta)\}$ has a finite limit for $|\theta| < b$ and by Theorem 2 of Kozakiewicz (1947) $H(x)$ is a proper D.F. and $\psi(\theta)$ its m.g.f.. Hence, since $\phi(\theta)$ is an m.g.f., $\gamma(0+) = 1$ and the continuity theorem for Laplace-Stieltjes transforms (Feller, 1971) shows that $G(x)$ is a proper D.F.. Thus $F_{n'}(x) \to G(x) + H(x) - U(x)$ on a dense subset of the real line and so is equal to $F(x)$, which is thus a proper D.F., and $\phi(\theta) = \int_{-\infty}^{\infty} e^{-\theta x} dF(x)$ $(|\theta| < b)$. This last relation uniquely determines $F(x)$ so all convergent subsequences of $\{F_n(x)\}$ converge to $F(x)$, thus completing the proof.

3. Limit theorems for Markov chains

We now consider the case C_2 of the introduction although the following remarks will apply to C_1 if $\mathscr{S}$ is substituted for $\mathscr{I}$. If $X_{n-1} = i$, $X_n = j (i, j \in \mathscr{I})$, let $m_{ij}(\theta) = E(e^{-\theta Y_n})$ $(n = 1, 2, \cdots)$ and if $X_0 = i$, write $m_i(\theta) = E(e^{-\theta Y_0})$; all these m.g.f.'s are assumed to exist finitely for $\theta \in I = [0, b]$ $(0 < b < \infty)$. Let $P(\theta) = [p_{ij} m_{ij}(\theta);\ i, j \in \mathscr{I}]$ and denote the i, j element of $(P(\theta))^n$ by $p_{ij}^{(n)}(\theta)$. Finally letting $\{p_i\}_{i \in \mathscr{I}}$ be the initial distribution we have

$$E(e^{-\theta S_n}; X_n = j) = \sum_{i \in \mathscr{I}} p_i m_i(\theta) p_{ij}^{(n)}(\theta) \qquad (j \in \mathscr{I}).$$

The limit $\nu(\theta) = 1/R(\theta) = \lim_{n \to \infty} (p_{ij}^{(n)}(\theta))^{1/n}$ is known to exist and is called the convergence norm of $P(\theta)$; see e.g. Vere-Jones (1967). It follows as in Kingman (1961) that $\nu(\theta)$ is super-convex in I and thus $R(\theta)$ is continuous (Rockafellar, 1970).

There is little loss of generality in assuming that $P(0)$, hence $P(\theta)$, is aperiodic. Our fundamental assumption is that $P(\theta)$ is $R(\theta)$-positive (Vere-Jones, 1967), which means that

$$R^n(\theta) p_{ij}^{(n)}(\theta) \to e_i(\theta) \pi_j(\theta) \quad , \quad (\theta \in I, n \to \infty, i, j \in \mathscr{I})$$

where the right-hand side is positive, $\sum_{j \in \mathscr{I}} e_j(\theta) \pi_j(\theta) = 1$ and $\{e_i(\theta)\}$ and $\{\pi_j(\theta)\}$ are right and left $R(\theta)$-invariant vectors of $P(\theta)$. We shall further assume that the convergence above is uniform with respect to $\theta \in I$, and say that $P(\theta)$ is *uniformly $R(\theta)$-positive.* Let $e_j = e_j(0)$, $\pi_j = \pi_j(0)$ and $R = R(0)$. For case C_1 we assume that $R = 1$, that is, $P(0)$ is positive recurrent and $\{\pi_j\}$ is the stationary distribution. Finally, for case C_2, let $0 < q_i = P(T_i < \infty)$.

It is easily seen that

$$E(e^{-\theta S_n} \mid n + m < T < \infty) = \frac{\sum_{i,j,k \in \mathscr{I}} p_i m_i(\theta) p_{ij}^{(n)}(\theta) p_{jk}^{(m)} q_k}{\sum_{i,k \in \mathscr{I}} p_i p_{ik}^{(n+m)} q_k}. \tag{2}$$

For the case $m = 0$ we have the following conditional weak law of large numbers (W.L.L.N.) and central limit theorem (C.L.T.).

Theorem 2. *Suppose that $P(\theta)$ is uniformly $R(\theta)$-positive and*

(i) $\sum_{j \in \mathscr{I}} \pi_j(\theta) q_i$ *is uniformly convergent and continuous on I;*

(ii) *either $p_i < K\pi_i$ $(0 < K < \infty)$ or $\sum_{i \in \mathscr{I}} p_i e_i < \infty$ and $\{e_i\}$ is bounded away from zero;*

(iii) *$R(\theta)$ has a continuous derivative on I and $0 < \mu = R'(0+)/R < \infty$.*

Then the conditioned sequence $\{S_n/n\}$, given $n < T < \infty$, converges in probability to μ.

Theorem 3. *In addition to the conditions of Theorem* 2, *suppose that $R(\theta)$ has a continuous second order derivative on I and that $0 < \sigma^2 = \mu^2 - R''(0+)/R < \infty$. Then the conditioned sequence $\{(S_n - n\mu)/(\sigma\sqrt{n})\}$, given $n < T < \infty$, converges in law to the standard normal D.F..*

Remark. If $\pi_j(\theta)$ is continuous $(j \in \mathscr{I})$ then the uniform convergence in (i) follows from Dini's Theorem.

Theorem 6.2 of Vere-Jones (1967) shows that if

(iv) $\sum_{j \in \mathscr{I}} \pi_j q_j < \infty$ and $\sum_{i \in \mathscr{I}} p_i |m_i(\theta)| \, |p_{ij}^{(n)}(\theta)| < K\pi_j$ $(0 < K < \infty; \theta \in I, j \in \mathscr{I})$

or if $\sum_{i,j \in \mathscr{I}} p_i |m_i(\theta)| \, |p_{ij}^{(n)}(\theta)| e_j < \infty$ and $\{e_i\}$ is bounded away from zero, then the right-hand side of (2) has a limit as $m \to \infty$. (We allow the possibility here that θ may be imaginary, which will be needed later.) Denoting this limit by $E(e^{-\theta S_n} \mid T = \infty)$ we have

$$E(e^{-\theta S_n} \mid T = \infty) = R^n \sum_{i,j \in \mathscr{I}} p_i m_i(\theta) p_{ij}^{(n)}(\theta) e_j / \sum_{i \in \mathscr{I}} p_i e_i \tag{3}$$

and the right-hand side is to be interpreted as giving the distribution of S_n given that absorption does not take place within the foreseeable future. In the case that the Y_n are non-negative, conditions (iv) and (ii) are equivalent.

The following theorems give the W.L.L.N. and C.L.T. for S_n given $T = \infty$.

Theorem 4. *Suppose that the conditions of Theorem* 2 *hold but with condition* (iv) *replacing* (ii) *and also that*

(v) $\sum_{j \in \mathscr{I}} \pi_j(\theta) e_j$ *is uniformly convergent and continuous on I.*

Then the conditioned sequence $\{S_n/n\}$, given $T = \infty$, converges in probability to μ.

Theorem 5. *Let the conditions of Theorem* 4 *and the additional requirement of Theorem* 3 *hold. Then the conditioned sequence $\{(S_n - n\mu)/(\sigma\sqrt{n})\}$, given $T = \infty$, converges in law to the standard normal D.F..*

It was essentially shown in Pakes (1973 b) that if $P(0)$ is R-positive and if conditions covered by (ii) and (iii) hold, then $P(X_n = j \mid T = n+m) \to u_j^{(m)} > 0$ ($n \to \infty$; $m = 1, 2 \cdots$; $j \in \mathscr{I}$) and $\sum_{j \in \mathscr{I}} u_j^{(m)} = 1$. Writing $g_i^{(n)} = P(T_i = n)$, it is clear that

$$E(e^{-\theta S_n} \mid T = n+m) = \frac{\sum_{i,j \in \mathscr{I}} p_i m_i(\theta) p_{ij}^{(n)}(\theta) g_j^{(m)}}{\sum_{i \in \mathscr{I}} p_i g_i^{(n+m)}}.$$

The following theorems give a W.L.L.N. and C.L.T. for S_n given $T = n+m$.

Theorem 6. *Under the conditions of Theorem* 2 *the conditioned sequence* $\{S_n/n\}$, *given* $T = n+m$, *converges in probability* ($n \to \infty$) *to* μ.

Theorem 7. *Under the conditions of Theorem* 3, *the conditioned sequence* $\{(S_n - n\mu)/(\sigma\sqrt{n})\}$ *given* $T = n+m$, *converges in law to the standard normal D.F.*.

Turning now to the case $\mathscr{I} = \mathscr{S}$ we have the following C.L.T. for positive-recurrent Markov chains; see Revesz (1968, p. 130) for a strong law of large numbers.

Theorem 8. *With* $R = 1$, *let* $P(0)$ *be uniformly* $R(\theta)$*-positive and let* (i) (*with* $q_j \equiv 1, j \in \mathscr{S}$), (iii) *and the extra assumption of Theorem* 3 *hold. Then* $\{(S_n - n\mu)/(\sigma\sqrt{n})\}$ *converges in law to the standard normal D.F.*.

As the proofs of Theorems 2–8 are similar, we shall prove Theorem 5 only. Since $\pi_k(\theta) > 0$ ($k \in \mathscr{I}$), then for a fixed $i \in \mathscr{I}$ there is a constant M such that $\delta_{ik} m_k(\theta) < M\pi_k(\theta)$ ($k \in \mathscr{I}$, $\theta \in I$). Hence

$$\begin{aligned} S_{ij}^{(n)}(\theta) &= m_i(\theta) p_{ij}^{(n)}(\theta) R^n(\theta) \leq M R^n(\theta) \sum_{k \in \mathscr{I}} \pi_k(\theta) p_{kj}^{(n)}(\theta) \\ &= M\pi_j(\theta) \qquad\qquad (j \in \mathscr{I}, \theta \in I). \end{aligned}$$

Thus condition (v) shows that for each $\varepsilon > 0$, there is a set $\mathscr{I}(\varepsilon) \subset \mathscr{I}$ such that $\mathscr{I} \setminus \mathscr{I}(\varepsilon)$ is finite and $\sum_{j \in \mathscr{I}(\varepsilon)} S_{ij}^{(n)}(\theta) e_j < \varepsilon$ for all $\theta \in I$ and $n = 1, 2, \cdots$. It then follows readily from Cauchy's criterion that $\sum_{j \in \mathscr{I}} S_{ij}^{(n)}(\theta) e_j$ is uniformly convergent for $\theta \in I$ as $n \to \infty$. Theorem 6.2 of Vere-Jones (1967) shows that the limit is $m_i(\theta) e_i(\theta) \sum_{j \in \mathscr{I}} \pi_j(\theta) e_j$. Hence we obtain

$$E(e^{-\theta S_n} \mid T_i = \infty) = (R/R(\theta))^n (m_i(\theta) e_i(\theta)/e_i) \sum_{j \in \mathscr{I}} \pi_j(\theta) e_j (1 + r_i^{(n)}(\theta)),$$

where $r_i^{(n)}(\theta) \to 0$ ($n \to \infty$) uniformly for $\theta \in I$. Uniform $R(\theta)$-positivity implies that $e_i(\theta)\pi_j(\theta)$ is continuous and positive on I, hence $e_i(\theta)$ is bounded on I, so by (v) the series $\sum_{j \in \mathscr{I}} e_i(\theta) \pi_j(\theta) e_j$ is uniformly convergent, hence continuous on I. Its limit as $\theta \downarrow 0$ is e_i. The conditions on $R(\theta)$ yield the Taylor expansion

$$R(\theta)/R = 1 + \mu\theta - [(\sigma^2 - \mu^2)/2]\theta^2(1 + o(1)) \ (\theta \downarrow 0),$$

and it follows easily that

$$E(e^{-\theta(S_n - n\mu)/\sigma\sqrt{n}} \mid T_i = \infty) \to e^{\theta^2/2} \ (n \to \infty; \theta \in I).$$

In the case that $p_k = \delta_{ik}$ ($k \in \mathscr{I}$), the theorem now follows from Theorem 1.

Denote the characteristic function of $(S_n - n\mu)/(\sigma\sqrt{n})$, given $T_i = \infty$ (resp. $T = \infty$) by $\psi_i^{(n)}(\zeta)$ (resp. $\psi^{(n)}(\zeta)$). It follows from (3) that

$$\psi^{(n)}(\zeta) = \sum_{i \in \mathscr{I}} p_i e_i \psi_i^{(n)}(\zeta) / \sum_{i \in \mathscr{I}} p_i e_i .$$

Condition (iv) implies that $\sum_{i \in \mathscr{I}} p_i e_i < \infty$ and we have shown that $\psi_i^{(n)}(\zeta) \to e^{-\zeta^2/2}$ $(n \to \infty)$. The theorem now follows on using dominated convergence and the continuity theorem for characteristic functions.

4. APPLICATIONS TO GALTON-WATSON PROCESSES

We now consider the well known Galton-Watson process (Harris, 1963), which is a Markov chain $\{X_n\}$ with $\mathscr{S} = \{0, 1, \cdots\}$ and whose transition matrix is defined by $\sum_{j \in \mathscr{S}} P_{ij} x^j = [A(x)]^i$ $(i \in \mathscr{S})$ and $A(x) = \sum_{j \in \mathscr{S}} a_j x^j$ is a probability generating function (p.g.f.) with $0 < a_0$, $a_0 + a_1 < 1$. We interpret X_n as the number of progeny born to the X_{n-1} parents which compose the $(n-1)$th generation $(n = 1, 2, \cdots)$. Clearly $\{0\}$ is an absorbing state, and as is shown in Harris (1963, p. 26) there exists an infinite subset $\mathscr{I} \subseteq \{1, 2, \cdots\}$ such that $[p_{ij}, i, j \in \mathscr{I}]$ is strictly sub-stochastic, $\mathscr{I}$ is irreducible and $p_{ij}^{(n)} = 0$ $(j \notin \mathscr{I};\ n = 1, 2, \cdots)$.

By taking $m_j(\theta) = m_{ij}(\theta) = e^{-j\theta}$ $(j \in \mathscr{I})$ we shall show that when $\alpha = A'(1-) < 1$, (which we assume without further comment) the conditions of Theorems 2–7 may be satisfied and thus obtain conditional limit theorems for $S_n = X_0 + \cdots + X_n$, the total progeny up to the nth generation.

Let $p(x) = [p_{ij} x^j;\ i, j \in \mathscr{I}]$ $(0 \leqq x \leqq 1)$ and denote the i, j element of $(p(x))^n$ by $P_{ij}^{(n)}(x)$, so that $P_{ij}^{(n)}(e^{-\theta}) = p_{ij}^{(n)}(\theta)$. Denote the least positive solution of $g = xA(g)$ by $g(x)$ and this is known to be the p.g.f. of $\sum_{n=0}^{\infty} X_n$ when $X_0 = 1$. Define the sequence of functions $\{\gamma_n \equiv \gamma_n(x, y)\}$ by $\gamma_0 = xy$ and $\gamma_{n+1} = xA(\gamma_n)$ $(n = 0, 1, \cdots;\ 0 \leqq x,\ y \leqq 1)$; it is shown in Pakes (1971 b) that $\gamma_n = E(y^{X_n} x^{Y_n} \mid X_0 = 1)$. The behaviour of this sequence is summarized in the following lemma whose proof may be found in Pakes (1971 b).

Lemma 1. *For* $0 \leqq y \leqq 1$, $\gamma_n \uparrow g(x)$ *(resp.* $\downarrow g(x)$*) uniformly for* $0 \leqq x \leqq 1$ *if* y *satisfies* $xy \leqq g(x)$ *(resp.* $xy \geqq g(x)$*). If* $A''(1-) < \infty$ *then* $(g(x) - \gamma_n)/u^n \to \Gamma(x, y)$ $(n \to \infty)$ *uniformly for* $0 \leqq x \leqq 1$*, where* $|\Gamma(x, y)| < \infty$ *and* $u \equiv u(x) = xA'(g(x))$.

Theorem 9. *Let* $\alpha < 1$ *and* $A''(1-) < \infty$. *Then*

$$[u(x)]^{-n} P_{ij}^{(n)}(x) \to i(g(x)/x)^{i-1} \Pi_j(x) \qquad (n \to \infty) \tag{4}$$

uniformly for $0 \leqq x \leqq 1$, *where* $\sum_{j \in \mathscr{I}} \Pi_j(x) y^j = [\Gamma(x, y) - \Gamma(x, 0)]/x$. *Furthermore*,

$$\sum_{j \in \mathscr{I}} \Pi_j(x) < \infty \text{ and is continuous in } 0 \leqq x \leqq 1; \tag{5}$$

also

$$\sum_{j \in \mathscr{I}} j \Pi_j(x) < \infty \text{ and is continuous in } 0 \leqq x \leqq 1. \tag{6}$$

Proof. For $0 \leqq y \leqq 1$ let $\Phi_i^{(n)}(x, y) = \sum_{j \in \mathscr{I}} P_{ij}^{(n)}(x) y^j$ $(n = 1, 2, \cdots)$, so that on setting $\phi_n(x) = \gamma_n(x, 0) = E(x^{Y_n};\ X_n = 0)$, we obtain

$$\Phi_i^{(n)}(x, y) = [\gamma_n(x, y)/x]^i - [\phi_n(x)/x]^i$$

$$= [(\gamma_n(x, y) - \phi_n(x))/x^i] \sum_{k=0}^{i-1} (\gamma_n(x, y))^k (\phi_n(x))^{i-k-1}.$$

Lemma 1 shows that $\Psi_i^{(n)}(x, y) = u^{-n}\Phi_i^{(n)}(x, y)$ converges to $i(g(x)/x)^{i-1}(\Gamma(x, y) - \Gamma(x, 0))/x$ uniformly for $0 \leqq x \leqq 1$. When $xy \leqq g(x)$ it follows from Lemma 1 that

$$\Psi_1^{(n)}(x, y) \leqq (g(x) - \phi_n(x))/(xu^n) \leqq g(x)/x \leqq 1, \quad (0 \leqq x \leqq 1)$$

with the penultimate inequality following from equation (12) of Pakes (1971 b). When $xy \geqq g(x)$ an argument of the type on p. 181 of Pakes (1971 b) shows that $\Psi_1^{(n)}(x, y) \leqq \Pi_{m=0}^{n}(1 + K\alpha^m) < \infty$, where $0 < K < \infty$ is a constant. Thus $\{\Psi_1^{(n)}(x, y)\}$ is uniformly bounded, in $0 \leqq x,\ y \leqq 1$ and it follows from the continuity theorem for p.g.f.'s (Feller, 1970, p. 281) that $u^{-n}P_{1j}^{(n)}(x) \to \Pi_j(x)\,(n \to \infty)$, where the limit is as given in the statement of the theorem. Furthermore the proof of the continuity theorem can be easily adapted to show that the uniform convergence of $\Psi_1^{(n)}(x, y)$ implies that of $u^{-n}P_{1j}^{(n)}(x)$. The same argument may now be applied to $\Psi_i^{(n)}(x, y)$ to complete the proof of (4), and (5) follows immediately by letting $y = 1$.

It is shown in Pakes (1971 b, Lemma 1) that $\gamma_{n+1}(x, y) = \gamma_n(x, A(xy))$ and this yields the functional equation

$$u(x)\Gamma(x, y) = \Gamma(x, A(xy)). \tag{7}$$

Letting $\Delta(x, y) = \Sigma_{j\in\mathscr{I}}\Pi_j(x)jy^{j-1}$, differentiation of (7) yields the functional equation

$$V(x)\Delta(x, y) = B(xy)\Delta(x, A(xy)),$$

where $B(x) = A'(x)/\alpha$, and $V(x) = B(g(x))$. An equation of this form arose in Pakes (1973 c) and (6) follows from the discussion therein.

Setting $e_i(\theta) = i(e^\theta g(e^{-\theta}))^{i-1}$ we have $e_i = i$ so that the second part of condition (ii) in Theorem 2 is satisfied if the initial distribution has a finite mean. Taking $I = [0, b]\,(0 < b < \infty)$, $P(\theta)$ is seen to be uniformly $R(\theta)$-positive with $R(\theta) = e^\theta/A'(g(e^{-\theta}))$, and under the conditions of Theorem 9 we have

$$\mu = 1 + A''(1-)/(\alpha(1-\alpha)).$$

Finally, let $\pi_j(\theta) = \Pi_j(e^{-\theta})$ and since $e_i(\theta)$ is continuous on I so is $\pi_j(\theta)$, since their product is. For the subcritical Galton-Watson process $q_j = 1$, so it follows from (5), (6) and Dini's theorem that conditions (i) and (v) are satisfied. Thus we see that all the requirements for Theorems 2, 4 and 6 are satisfied and if $A'''(1) < \infty$ then those of Theorems 3, 5 and 7 are satisfied with

$$\sigma^2 = \frac{A'''(1) + (1+\alpha)A''(1)}{\alpha(1-\alpha)^2} - \frac{(1-2\alpha)(A''(1))^2}{\alpha^2(1-\alpha)^3}.$$

Theorems 2, 3, 6 and 7 were obtained by an alternative means in Pakes (1971b) for the case $p_j = \delta_{ij}\ (j \in \mathscr{I})$; note that the final sentence of §3 of this reference is false.

We shall now obtain results to show that the requirement of a finite mean for the initial distribution cannot in general be relaxed. This, of course, is not surprising in view of the results of Rubin & Vere-Jones (1968), for example. Let $\tau(x) = \Sigma_{i\in\mathscr{I}}p_ix^i$ and

$$1 - \tau(1-x) = x^\delta L(x) \quad , \quad (0 < \delta < 1) \tag{8}$$

where $L(x)$ is slowly varying at the origin. The following theorem shows that when (8) holds the norming constants required to obtain a limiting distribution for S_n, given $X_n > 0$, increase geometrically, rather than linearly as for the conditional W.L.L.N.. We shall also see that, in contrast to our earlier results, varying the conditioning events varies the limit distribution.

Writing $g_n(x) = \gamma_n(x,1)$, $A_1 = A(0)$, $A_{n+1} = A(A_n)$ $(n = 0,1,\cdots)$, it is clear that

$$E(x^{S_n} \mid X_n > 0) = [\tau(g_n(x)) - \tau(\phi_n(x))]/(1 - \tau(A_n)). \tag{9}$$

Theorem 10. *Let* $A''(1-) < \infty$ *and* (8) *hold for the subcritical* $(\alpha < 1)$ *Galton-Watson process. Then the conditional sequence* $\{\alpha^n S_n\}$, *given* $X_n > 0$, *converges in law to the distribution with density*

$$f_v(t) = \begin{cases} (v\delta/\Gamma(1-\delta))(vt)^{-1-\delta}(1-e^{-vt}) & (t > 0) \\ 0 & (t < 0), \end{cases} \tag{10}$$

where $v = (1-\alpha)/c$ *and* $c = \lim_{n\to\infty} \alpha^n/(1 - A_n)$.

Proof. Denoting the left-hand side of (8) by $T(x)$ and writing $\theta_n = \exp(-\theta\alpha^n)$, (9) becomes

$$E(\exp(-\theta\alpha^n S_n) \mid X_n > 0) = \frac{T(1-\phi_n(\theta_n)) - T(1-g_n(\theta_n))}{T(1-A_n)}. \tag{11}$$

Kolmogorov's theorem (Harris, 1963, p. 18) shows that c exists and $0 < c < \infty$. Since $g'(1) = (1-\alpha)^{-1}$, the mean value theorem shows that $(1 - g(\theta_n))/(1-A_n) \to c\theta/(1-\alpha)$. Theorem 1 and Lemma 5 of Pakes (1971b) show that $g(\theta_n) - \phi_n(\theta_n) \sim \alpha^n/c$ and $g_n(\theta_n) - g(\theta_n) = o(\alpha^n)$. Hence we obtain $(1-\phi_n(\theta_n))/(1-A_n) \to 1 + c\theta/(1-\alpha)$ and $(1-g_n(\theta_n))/(1-A_n) \to c\theta/(1-\alpha)$. Now $T(\lambda x)/T(x) \to \lambda^\delta$ $(\lambda \downarrow 0)$ uniformly in subsets of the form $0 < \varepsilon \leqq \lambda \leqq \delta < \infty$. Thus the right-hand side of (11) $\to [1 + c\theta/(1-\alpha)]^\delta - [c\theta/(1-\alpha)]^\delta$ and the theorem now follows from the continuity theorem for Laplace-Stieltjes transforms (L.S.T.'s) and the easily verified fact that $\int_0^\infty e^{-\theta t} f_v(t)dt = (1 + \theta/v)^\delta - (\theta/v)^\delta$.

Still assuming that (8) holds, the following theorem shows that Theorem 4 fails in the sense that $P(S_n = j \mid X_{n+m} > 0) \to 0 (m \to \infty;\ j \in \mathscr{I})$. Indeed, we have the stronger result that $P(X_n = j \mid X_{n+m} > 0) \to 0$ $(m \to \infty)$, which should be contrasted with the situation when $\phi'(1) < \infty$ discussed in Seneta & Vere-Jones (1966).

Theorem 11. *Under the conditions of Theorem* 10 *the conditional sequence* $\{\alpha^m X_n\}$ *(resp.* $\{\alpha^m S_n\}$*), given* $X_{n+m} > 0$, *converges in law* $(m \to \infty)$ *to the distribution with density* $f_v(t)$, *where* $v = c^{-1}$ *(resp.* $v = (1-\alpha)/\{c(\alpha^{-n} - \alpha)\}$*) and* $f_v(\cdot)$ *is given by* (10).

Proof. It is well known that $E(x^{X_n} \mid X_0 = i) = (A_n(x))^i$ where $A_1(x) = A(x)$ and $A_{n+1}(x) = A(A_n(x))$ $(n = 1,2,\cdots)$ so that

$$E(x^{X_n} \mid X_{n+m} > 0) = \frac{T(1-A_n(xA_m)) - T(1-A_n(x))}{T(1-A_{n+m})}.$$

Kolmogorov's Theorem says that $1 - A_{n+m} \sim \alpha^{n+m}/c$ and clearly $1 - A_n(\theta_m) \sim \theta\alpha^{n+m} (m \to \infty)$. Furthermore $1 - A_m\theta_m \sim (\theta + c^{-1})\alpha^m$, hence $1 - A_n(\theta_m A_m) \sim (\theta + c^{-1})\alpha^{n+m}$ $(m \to \infty)$ and the assertion regarding $\{\alpha^m X_n\}$ now follows as in Theorem 10.

For the second part we observe that

$$E(S_n \mid X_{n+m} > 0) = \frac{T(1-\gamma_n(x,A_m)) - T(1-g_n(x))}{T(1-A_{n+m})}.$$

It is easily shown that $1 - g_n(\theta_m) \sim \theta\alpha^m(1-\alpha^{n+1})/(1-\alpha)$ and Taylor's Theorem for several variables yields

$$1 - \gamma_n(\theta_m, A_m) \sim \alpha^n(1 - A_m) + (1-\theta_m)(1-\alpha^{n+1})/(1-\alpha)$$
$$\sim c^{-1}\alpha^{n+m} + \theta\alpha^m(1-\alpha^{n+1})/(1-\alpha) \qquad (m \to \infty).$$

The Theorem readily follows.

Letting T be the time to extinction of the Galton-Watson process, we shall now see how Theorem 6 fails when (8) holds. It should be noted that, given $T = n + m$, X_n converges in law to a discrete distribution which has moments of all orders; see for example Papangelou (1968).

Theorem 12. *Under the conditions of Theorem* 10 *the conditional sequence* $\{\alpha^n S_n\}$, *given* $T = n + m$, *converges in law* $(n \to \infty)$ *to a distribution with density*

$$h_v(t) = \frac{v\delta(vt)^{-1-\delta}(e^{-\alpha vt} - e^{-vt})}{(1-\alpha^\delta)\Gamma(1-\delta)}, \qquad (t \geqq 0)$$

where $v = c^{-1}\alpha^{m-2}(1-\alpha)$.

Proof. It is readily seen that

$$E(x^{S_n} \mid T = n+m) = \frac{T(1-\gamma_n(x, A_{m-1})) - T(1-\gamma_n(x, A_m))}{T(1-A_{n+m-1}) - T(1-A_{n+m})}.$$

It follows from Lemma 1 that $g(\theta_n) - \gamma_n(\theta_n, A_m) \sim \alpha^n\Gamma(1, A_m)$ and $\Gamma(1, A_m) = \lim_{n\to\infty}(1-\gamma_n(1,A_m))\alpha^n$ $= \lim_{n\to\infty}(1 - A_n(A_m))/\alpha^n = c^{-1}\alpha^m$. Hence $1 - \gamma_n(\theta_n, A_m) \sim \alpha^n(\theta/(1-\alpha) + c^{-1}\alpha^m)$, and as in previous proofs we find that

$$E(e^{-\theta\alpha^n S_n} \mid T = n+m) \to ((1+\zeta)^\delta - (\alpha+\zeta)^\delta)/(1-\alpha^\delta) \qquad (n \to \infty)$$

where $\zeta = c\theta\alpha^{-m+2}/(1-\alpha)$, and the proof is now easily completed.

We now consider a branching process in which objects reproduce as in a Galton-Watson process and which is also subject to an independent immigration component at each generation. Let $B(x)$ $(0 < B(0) < 1)$ be the p.g.f. of the number of immigrants per generation. Taking Z_n to be the number of objects making up the nth generation $(n = 0, 1, \cdots)$, it is known that $\{Z_n\}$ is a Markov chain with state space $\{0, 1, \cdots\}$, and that the transition probabilities are given by $\sum_{j=0}^{\infty} p_{ij}x_j = B(x)(A(x))^i$ $(i = 0, 1, \cdots)$. It is further known that there is a subset, $\mathscr{S}$, of the non-negative integers which contains $\{0\}$ and is an irreducible and aperiodic set of states. If $\beta = B'(1-) < \infty$ and $\alpha < 1$ then $\{Z_n\}$ is ergodic; see Seneta (1969) for a summary and references for these matters.

Again, taking $m_j(\theta) = m_{ij}(\theta) = e^{-j\theta}$, we shall be considering $S_n = Z_0 + \cdots + Z_n$, the total number of objects which have existed up to the nth generation. Assuming that $A''(1-) < \infty$, then it is essentially shown in Pakes (1973c) that $p_{ij}^{(n)}(\theta)$ is uniformly $R(\theta)$-positive with $R(\theta) = 1/B(g(\theta))$, $e_i(\theta) = (g(\theta))^i$, and $\pi_j(\theta)$ $(i, j \in \mathscr{S})$ satisfying the requirements of Theorem 8 when $B''(1-) < \infty$. This gives the following C.L.T. which was first obtained by Pakes (1971a) as a consequence of a C.L.T. for triangular arrays.

Theorem 13. *If* $A''(1-)$, $B''(1-) < \infty$ *for the subcritical* $(\alpha < 1)$ *Galton-Watson process with immigration, then* $(S_n - \mu n)/(\sigma\sqrt{n})$ *converges in law to the standard normal distribution, where* $\mu = \beta/(1-\alpha)$, $\sigma^2 = \beta\sigma_A^2/(1-\alpha)^3 + \sigma_B^2/(1-\alpha)^2$, *and* σ_A^2 *and* σ_B^2 *are the variances of the offspring and immigration distributions, respectively.*

5. CONTINUOUS TIME PROCESSES

Let us now consider a Markov process $\{X(t)\}$ having a countable state space $\mathscr{S}$, let $p_{ij}(t) = P(X(t+\tau) = j \mid X(\tau) = i)$ $(i, j \in \mathscr{S}; t, \tau \geqq 0)$ and always assume the transition matrix $[p_{ij}(t)]$ is standard (Chung, 1967). For the time being we take $\mathscr{I}$ to be an infinite irreducible proper subset of $\mathscr{S}$ such that $\sum_{j\in\mathscr{I}} p_{ij}(t) < 1$ for some $i \in \mathscr{I}$ and $p_{ij}(t) \equiv 0$ for $i \in \mathscr{A} = \mathscr{S} \setminus \mathscr{I}$ and $j \in \mathscr{I}$. Thus $\mathscr{A}$ is an absorbing set. Let $\{p_i\}$ $(i \in \mathscr{I})$ be the initial distribution of the process.

The matrix $P(t) = [p_{ij}(t); i, j \in \mathscr{I}]$ is said to have decay parameter λ if this is the common limit of the functions $-t^{-1}\log p_{ij}(t)$ $(t \to \infty)$ and to be λ-positive if the limits of $e^{\lambda t}p_{ij}(t)$ exist and are finite and positive. When this is the case these limits are of the form $e_i\pi_j$ with $\sum_{i\in\mathscr{I}} e_i\pi_i = 1$ and satisfy (uniquely up to constant multiples) the equations:

$$\sum_{i\in\mathscr{I}} \pi_i p_{ij}(t) = \pi_j e^{-\lambda t}, \quad \sum_{j\in\mathscr{I}} p_{ij}(t)e_j = e_i e^{-\lambda t}. \tag{12}$$

See Kingman (1963a) for a thorough exposition of these matters. Let $T_i = \inf\{t \mid X(t) \in \mathscr{A}, X(0) = i \in \mathscr{I}\}$ be the time to absorption, and denote its D.F. by $F_i(t) = \sum_{j\in\mathscr{A}} p_{ij}(t)$. Finally we define $q_i = F_i(\infty-) = P(T_i < \infty)$ and for a general initial distribution denote the time to absorption by T, whose D.F., $F(t)$, is then seen to be $\sum_{i\in\mathscr{I}} p_i F_i(t)$.

It is not difficult to formulate and prove analogues of the discrete time conditional limit theorems discussed by Seneta & Vere-Jones (1966). We shall state two such results and refer the reader to Vere-Jones (1969) where the first is stated and proved (in the case that $p_j = \delta_{ij}$).

Theorem 14. *If* $P(t)$ *is* λ*-positive and either of the following conditions hold*:

(a) $$\sum_{j\in\mathscr{I}} \pi_j q_j < \infty \text{ and } p_j < K\pi_j \quad (0 < K < \infty, j \in \mathscr{I}); \text{ or}$$

(b) $$\sum_{i\in\mathscr{I}} p_i e_i < \infty \text{ and } q_i < K' e_i \quad (0 < K' < \infty, i \in \mathscr{I}),$$

then

$$\lim_{t\to\infty} P(X(t) = j \mid t < T < \infty) = \pi_j q_j / \sum_{j\in\mathscr{I}} \pi_j q_j \qquad (j \in \mathscr{I}),$$

and

$$\lim_{t\to\infty}\lim_{\tau\to\infty} P(X(t) = j \mid t+\tau < T < \infty) = e_j\pi_j \qquad (j \in \mathscr{I}).$$

Note that if the initial distribution is concentrated on a finite subset of $\mathscr{I}$ then the first part of (a) is necessary and sufficient for the Theorem. A sufficient condition for the second part of (b) is that $\{e_i\}$ be bounded away from zero; see Seneta & Vere-Jones (1966).

It is easily seen that

$$P(X(t) = j \mid t+\tau < T \leqq t+\tau+h) = \frac{\sum_{i\in\mathscr{I}} p_i p_{ij}(t)(F_j(\tau+h) - F_j(\tau))}{F(t+\tau+h) - F(t+\tau)}. \tag{13}$$

Since $[p_{ij}(t)]$ is standard, the limits $q_{ii} = \lim_{h\downarrow 0}(p_{ii}(h) - 1)/h$ exist and we assume them to be finite $(i \in \mathscr{S})$. It then follows from Chung (1967, p. 135) that T_i has a density function given by

$$f_i(t) = \sum_{j \in \mathscr{A}} p'_{ij}(t) \tag{14}$$

and continuous for each $t > 0$. This follows from the uniform convergence of the last series in $t \geqq \delta > 0$. Monotone convergence then shows that $f(t) = \sum_{i \in \mathscr{I}} p_i f_i(t)$ is a density, namely that of T. Thus we now see that the left-hand side of (13) tends to a limit as $h \to 0$ and we shall denote this by

$$P(X(t) = j \mid T = t + \tau) = \sum_{i \in \mathscr{I}} p_i p_{ij}(t) f_j(\tau)/f(t+\tau). \tag{15}$$

Alternative expressions for $f(t)$ and $f_i(t)$ are obtained by observing that $P(t < T_i < \infty) = \sum_{j \in \mathscr{I}} p_{ij}(t) q_j$ and that, by the cited results of Chung, the series is term-wise differentiable, thus yielding

$$f_i(t) = - \sum_{j \in \mathscr{I}} p'_{ij}(t) q_j. \tag{16}$$

If $\sum_{i \in \mathscr{I}} p_i e_i < \infty$, it follows from λ-positivity, Theorem 6.1 of Vere-Jones and a Croftian Theorem (see e.g. Vere-Jones, 1969, p. 74 and Kingman, 1963 b) that the series in (15) $\sim e^{-\lambda t}(\sum_{i \in \mathscr{I}} p_i e_i) f_j(\tau) \pi_j$ $(t \to \infty)$. To determine the asymptotic behaviour of $f(t+\tau)$ for large t we observe that (14), Fubini's Theorem and the cited theorems of Chung (1967) yield $f_i(t+\tau) = \sum_{k \in \mathscr{I}} p_{ik}(t) f_k(\tau)$. Now under conditions (c) or (d) below it follows from Theorem 6.2 of Vere-Jones (1967) and a Croftian Theorem that $f(t+\tau) \sim e^{-\lambda t}(\sum_{i \in \mathscr{I}} p_i e_i)(\sum_{j \in \mathscr{I}} \pi_j f_j(\tau))$ $(t \to \infty)$. In order that the Croftian Theorem be applicable we need to check that $f(t+\tau)$ is continuous for $t > 0$ and we show this for condition (c). We have for $h > 0$

$$\begin{aligned}
\left| f(t+\tau+h) - f(t+\tau) \right| &\leqq \sum_{i\,k \in \mathscr{I}} p_i \left| p_{ik}(t+h) - p_{ik}(t) \right| f_k(\tau) \\
&\leqq \sum_{i\,j\,k \in \mathscr{I}} p_i p_{ij}(t) \left| p_{jk}(h) - \delta_{jk} \right| f_k(\tau) \\
&\leqq K e^{-\lambda t} \sum_{j\,k \in \mathscr{I}} \pi_j \left| p_{jk}(h) - \delta_{jk} \right| f_k(\tau), \text{ on using (c) and (12)} \\
&\leqq K\Big[\sum_{k \in \mathscr{I}} (1 - p_{kk}(h)) \pi_k f_k(\tau) + \sum_{j,k \in \mathscr{I}} \pi_j p_{jk}(h) f_k(\tau) - \sum_{k \in \mathscr{I}} \pi_k p_{kk}(h) f_k(\tau) \Big].
\end{aligned}$$

Dominated convergence and (c) show that the first sum $\to 0$ $(h \downarrow 0)$ and on using (12) in the second sum it is clear that the difference of the other two series $\to 0$. The case $h \uparrow 0$ is handled similarly. This completes the proof of

Theorem 15. *If $P(t)$ is λ-positive and if, for fixed $\tau > 0$, either*

(c) $\sum_{j \in \mathscr{I}} \pi_j f_j(\tau) < \infty$ *and* $p_i < K\pi_i$ $(0 < K < \infty;\ i \in \mathscr{I})$; *or*

(d) $\sum_{i \in \mathscr{I}} p_i e_i < \infty$ *and* $f_i(\tau) < K' e_i$ $(0 < K' < \infty;\ i \in \mathscr{I})$,

then

$$\lim_{t \to \infty} P(X(t) = j \mid T = t + \tau) = u_j(\tau) \quad (j \in \mathscr{I}), \tag{17}$$

where

$$u_j(\tau) = \pi_j f_j(\tau) / \sum_{j \in \mathscr{I}} \pi_j f_j(\tau) \text{ and } \sum_{j \in \mathscr{I}} u_j(\tau) = 1.$$

This is the continuous time analogue of the Theorem mentioned immediately after Theorem 5. By imposing extra assumptions we shall find alternative forms of $u_j(\tau)$. Now the limits $q_{ij} = \lim_{t\downarrow 0} p_{ij}(t)/t\ (i \neq j)$ exist, and if the forward equations, $p'_{ij}(\tau) = \sum_{l\in\mathscr{I}} p_{il}(\tau)q_{lj}$, hold, and if $\sum_{j\in\mathscr{A}} q_{ij} = 0\ (i \in \mathscr{A})$, then from Fubini's Theorem, (12) and (14) we obtain

$$\sum_{i\in\mathscr{I}} \pi_i f_i(\tau) = e^{-\lambda t} \sum_{l\in\mathscr{I}, k\in\mathscr{A}} \pi_l q_{lk}. \tag{18}$$

If in addition $\sum_{j\in\mathscr{S}} q_{ij} = 0$, so that the backward equations are satisfied, then on defining $q_i = 1\ (i \in \mathscr{A})$ and $Q_{ij}(t) = q_j p_{ij}(t)/q_i\ (i,j \in \mathscr{S})$, it is clear that $[Q_{ij}(t)]$ is a standard transition matrix on $\mathscr{S}$ which also satisfies the appropriate backward equations. This fact implies (Chung, 1967, p. 247) that $\sum_{k\in\mathscr{A}} q_{lk} = -\sum_{k\in\mathscr{I}} q_{lk}q_k\ (l \in \mathscr{S})$. Substituting this into (18) and using Theorem 5 of Vere-Jones (1969) finally yields $\sum_{j\in\mathscr{I}} \pi_j f_j(\tau) = \lambda e^{-\lambda\tau} \sum_{k\in\mathscr{I}} \pi_k q_k$. Thus we have proved the following

Corollary. If $P(t)$ is λ-positive, the forward equations hold and $\sum_{j\in\mathscr{A}} q_{ij} = 0\ (i \in \mathscr{A})$, then a sufficient condition for (17) *is*

(e) $$\sum_{i\in\mathscr{I}} \pi_i q_{ij} < \infty\ (j \in \mathscr{A}) \text{ and } p_i < K\pi_i,$$

and in this case

$$u_j(\tau) = e^{\lambda\tau}\pi_j f_j(\tau) / \Big(\sum_{i\in\mathscr{I}, j\in\mathscr{A}} \pi_i q_{ij} \Big). \tag{19}$$

If, in addition, $\sum_{j\in\mathscr{S}} q_{ij} = 0$ then a sufficient condition replacing (e) *is condition* (a) *of Theorem* 14, *and in this case the denominator of* (19) *becomes $\lambda \sum_{i\in\mathscr{I}} \pi_i q_i$.*

Returning to (15) and starting from the expression

$$f_i(t+\tau) = -\sum_{k\ j\in\mathscr{I}} p'_{ik}(\tau)p_{kj}(t)q_j,$$

which follows from (16) and Chung (1967, p. 135) it may be seen that condition (b) of Theorem 14 can replace condition (a) in the Corollary. The demonstration of this uses the fact that (12) and the backward equation imply that

$$\sum_{j\in\mathscr{I}} q_{ij}e_j = -\lambda e_i \qquad (i \in \mathscr{I}),$$

the proof of which parallels that of Theorem 5 of Vere-Jones (1969).

We now associate with the sample path of $X(\cdot)$ over the interval $[\tau, \tau+t]$ a homogeneous additive continuous functional, $S_{\tau,\tau+t}$, whose distribution satisfies the following conditions:

(I) The quantities $p_{ij}(\theta,t) = E(e^{-\theta S_{\tau,\tau+t}}; X(\tau+t) = j \mid X(\tau) = i)$ exist for $t, \tau > 0$, $i, j \in \mathscr{S}$ and $\theta \in I = [0,\delta]$ where δ is a finite positive constant, $p_{ij}(\theta,0) = \delta_{ij}$ and $p_{ij}(0+,t) = p_{ij}(t)$;

(II) $\mu_i(\theta,t) = \sum_{j\in\mathscr{S}} p_{ij}(\theta,t) < M < \infty$ for $i \in \mathscr{S}$, $\theta \in I$ and $t > 0$;

(III) $$p_{ij}(\theta, t+\tau) = \sum_{k\in\mathscr{S}} p_{ik}(\theta,t)p_{kj}(\theta,\tau) \qquad (i,j \in \mathscr{S};\ t,\tau \geqq 0);$$

(IV) $$p_{ij}(\theta, 0+) = \delta_{ij} \text{ and } p_{ii}(\theta,t),\ \mu_i(\theta,t) \to 1 \qquad (t \downarrow 0)$$

uniformly for $\theta \in I$.

Condition (II) and the uniform convergence in (IV) are automatically satisfied in the case of non-negative functionals. We also remark that properties of integral functionals for irreducible $\mathscr{S}$ have been studied by Puri (1971).

Writing $P(\theta,t) = [p_{ij}(\theta,t);\, i,j \in \mathscr{I}]$ we shall assume that $P(\theta,t)$ is uniformly $\lambda(\theta)$-positive, that is, $e^{t\lambda(\theta)}p_{ij}(\theta,t) \to e_i(\theta)\pi_j(\theta) > 0$ $(t \to \infty)$ for each $i,j \in \mathscr{I}$ and uniformly for $\theta \in I$, where $\lambda(\theta)$ is the convergence norm of $P(\theta,t)$. We can obtain limit theorems for $S_t = S_{0,t}$ analogous to Theorems 2–8. Indeed, we have the following analogues of Theorems 2 and 3.

Theorem 16. *Let $P(\theta,t)$ be uniformly $\lambda(\theta)$-positive and the analogues (which take precisely the form) of conditions* (i) *and* (ii) *of Theorem* 2 *hold. Suppose also that $\lambda(\theta)$ has a continuous derivative on I with $0 < \mu = \lambda'(0) < \infty$. Then the conditioned functional S_t/t, given $t < T < \infty$, converges in probability to μ.*

Theorem 17. *If the conditions of Theorem* 16 *hold and if $\lambda(\theta)$ has a continuous second derivative on I with $0 < \sigma^2 = -\lambda''(0) < \infty$, then $(S_t - \mu t)/(\sigma\sqrt{t})$, given $t < T < \infty$, converges in law to the standard normal D.F..*

Proof. We shall prove only Theorem 16. It suffices to show that

$$\psi_i(\theta,t) = E(e^{-\theta S_t/t}\,|\,t < T_i < \infty) \to e^{-\mu\theta} \quad (t \to \infty;\, i \in \mathscr{I}), \tag{20}$$

for then application of Theorem 1 to all subsequences of $\psi_i(\theta,t)$ establishes the Theorem when the initial distribution is concentrated at a single point in $\mathscr{I}$. The general case may then be obtained by using characteristic functions as in the proof of Theorem 5.

Observe that (III) implies that $p_{ij}^{(n)}(\theta) = p_{ij}(\theta, nh)$ $(h > 0)$ satisfies the conditions of Theorem 2 with $m_i(\theta) \equiv 1$ and $R(\theta) = e^{h\lambda(\theta)}$, so that (20) holds with $t = nh$ $(n \to \infty;\, h > 0)$. A Croftian Theorem will then yield (20) once we show that $\psi_i(\theta,t)$ is continuous at each $t > 0$. To see this, note that $\psi_i(\theta,t) = \phi_i(\theta/t, t)$ where

$$\phi_i(\theta,t) = \sum_{j \in \mathscr{I}} p_{ij}(\theta,t)q_j \Big/ \sum_{j \in \mathscr{I}} p_{ij}(t)q_j. \tag{21}$$

The denominator in (21) is continuous for each $t > 0$ and for $t,\, h > 0$

$$\Big|\sum_{j \in \mathscr{I}} (p_{ij}(\theta,t+h) - p_{ij}(\theta,t))q_j\Big| \leqq \sum_{j \in \mathscr{I}} |p_{ij}(\theta,t+h) - p_{ij}(\theta,t)|$$

$$= \sum_{j \in \mathscr{I}} \Big|\sum_{k \in \mathscr{S}} [p_{ik}(\theta,h) - \delta_{ik}]p_{kj}(\theta,t)\Big| \leqq M[\mu_i(\theta,h) - 2p_{ii}(\theta,h) + 1],$$

and by (IV) the last term $\to 0$ $(h \downarrow 0)$ uniformly for $\theta \in I$. A similar procedure applies when $h \uparrow 0$. Thus $\phi_i(\theta,t)$ is continuous at $t > 0$ uniformly for $\theta \in I$, and the continuity of $\psi_i(\theta,t)$ now follows easily.

It is clear that

$$E(e^{-\theta S_t}\,|\,t+\tau < T < \infty) = \frac{\sum_{i,j\,k \in \mathscr{I}} p_i p_{ij}(\theta,t)p_{jk}(\tau)q_k}{\sum_{i,j \in \mathscr{I}} p_i p_{ij}(t+\tau)q_j} \tag{22}$$

and it follows by the M-test and a simple modification of Theorem 1 of Chung (1967, p. 120) that the left-hand side of (22) is continuous for each $\tau > 0$. It then follows from $\lambda(\theta)$-positivity that if the following holds:

$$\sum_{j\in\mathscr{I}} \pi_j q_j < \infty \text{ and } \sum_{i\in\mathscr{I}} p_i \left| p_{ij}(\theta,t) \right| < K\pi_j; \text{ or if} \tag{23}$$

$$\sum_{i,j\in\mathscr{I}} p_i \left| p_{ij}(\theta,t) \right| e_j < \infty \text{ and } \{e_i\} \text{ is bounded away from zero;}$$

then the right-hand side of (22) has a limit, $E(e^{-\theta S_t} \mid T = \infty)$, as $\tau \to \infty$, given by

$$E(e^{-\theta S_t} \mid T = \infty) = \frac{e^{\lambda t} \sum_{i,j\in\mathscr{I}} p_i p_{ij}(\theta,t) e_j}{\sum_{i\in\mathscr{I}} p_i e_i}.$$

We allow the possibility of θ being imaginary in (23) and note that if $S_t \geqq 0$ then (23) is equivalent to the "analogue" conditions referred to in Theorem 16. The following are analogues of Theorems 4 and 5.

Theorem 18. *Suppose that the conditions of Theorem* 16 *hold but with the "analogue" conditions therein being replaced by* (23) *and also that*

(f) $\sum_{j\in\mathscr{I}} \pi_j(\theta) e_j$ *is uniformly convergent and continuous in* I.

Then S_t/t, *given* $T = \infty$, *converges in probability to* μ.

Theorem 19. *If the conditions of Theorem* 16 *and the additional assumption of Theorem* 17 *hold, then* $(S_t - \mu t)/(\sigma\sqrt{t})$, *given* $T = \infty$, *converges in law to the standard normal D.F.*.

Proof. The proof is similar to that of Theorems 5 and 16. It is only necessary to see that $\sigma_i(\theta,t) = \sum_{j\in\mathscr{I}} p_{ij}(\theta,t)e_j$ is continuous at $t > 0$ uniformly for $\theta \in I$. To see this first note that

$$p_{ij}(\theta,t) \leqq (\pi_i(\theta))^{-1} \sum_{i\in\mathscr{I}} \pi_i(\theta) p_{ij}(\theta,t) = e^{-\lambda(\theta)t} \pi_j(\theta)/\pi_i(\theta).$$

Then, for $t, h > 0$,

$$\begin{aligned} \left| \sigma_i(\theta, t+h) - \sigma_i(\theta,t) \right| &\leqq \sum_{k,j\in\mathscr{I}} p_{ik}(\theta,t) \left| p_{kj}(\theta,h) - \delta_{kj} \right| e_j \\ &\leqq (\pi_i(\theta))^{-1} \sum_{k,j\in\mathscr{I}} \pi_k(\theta) \left| p_{kj}(\theta,h) - \delta_{kj} \right| e_j \\ &= (\pi_i(\theta))^{-1} \sum_{j\in\mathscr{I}} [1 + e^{-h\lambda(\theta)} - 2p_{jj}(\theta,h)] \pi_j(\theta) e_j. \end{aligned}$$

Now $\pi_i(\theta)$ is positive for each $\theta \in I$, so the coefficient of the last series is uniformly bounded and by (IV) and (f) the series can be made uniformly small.

We now consider the quantity

$$E(e^{-\theta S_t} \mid T = t + \tau) = \frac{\sum_{i,j\in\mathscr{I}} p_i p_{ij}(\theta,t) f_j(\tau)}{f(t+\tau)}$$

which may be arrived at in the same way as (15). Similar proofs as those employed above then yield the following analogue of Theorems 6 and 7.

Theorem 20. *Let* $P(\theta,t)$ *be uniformly* $\lambda(\theta)$*-positive,* $\lambda(\theta)$ *satisfy the condition of Theorem* 16, *condition* (c) *or* (d) *of Theorem* 15 *hold and* $\sum_{j\in\mathscr{I}} \pi_j(\theta) f_j(\tau)$ *converge uniformly and be continuous in* I. *Then* S_t/t, *given* $T = t + \tau$, *converges in probability to* μ. *If the* *additional assumptions of Theorem 17 are satisfied,* $(S_t - \mu t)/(\sigma\sqrt{t})$ *given* $T = t + \tau$, *converges in law to the standard normal D.F..*

Finally, we consider the case where $\mathscr{I} = \mathscr{S}$ and $\lambda = 0$ so that $\mathscr{S}$ is a positive recurrent class of $X(t)$. Still assuming that (I)–(IV) hold we obtain the following result.

Theorem 21. *If* $P(\theta, t)$ *is uniformly* $\lambda(\theta)$*-positive,* $\sum_{j\in\mathscr{S}}\pi_j(\theta)$ *converges uniformly and is continuous in I and if* $\lambda(\theta)$ *is as in Theorem 16 then* S_t/t *converges in probability to* μ. *If* $\lambda(\theta)$ *is as in Theorem* 17, *then* $(S_t - \mu t)/(\sigma\sqrt{t})$ *converges in law to the standard normal D.F..*

When $\mathscr{S}$ is finite a stronger form of the C.L.T. part of this Theorem may be found in Fukushima & Hitsuda (1967).

6. Applications to Markov branching processes

We consider the Markov branching process (M.B.P.) in which the lifetime of each object has the exponential D.F. $1 - e^{-\nu t}$ $(t \geqq 0)$ and at the end of its life each object reproduces independently of other objects present. The number of offspring per individual has p.g.f. $A(x)$ which satisfies the conditions given in §4 and, in addition, $a_1 = 0$. See Harris (1963, Chap. 5) for an account of this process. Letting $X(t)$ denote the number of objects at time t, it is known that $X(t)$ is a Markov process whose state space is the set of non-negative integers. The state $\{0\}$ is absorbing with $q_j = q^i$ where q is the least positive solution of $a(q) = 0$ where $a(x) = \nu(A(x) - x)$ and $q = 1$ iff $a = a'(1) \leqq 0$. We consider the subcritical case $a < 0$. Furthermore there is an infinite subset, $\mathscr{I}$, of the positive integers which is irreducible and transient and we can take $\mathscr{S} = \mathscr{I} \cup \{0\}$.

By employing the construction given in Harris (1963, Chap. 6), $X(t)$ becomes, with probability one, a step function with at most a finite number of jumps in any finite interval. Thus $Y(t) = \int_0^t X(\tau)d\tau$ is a well-defined stochastic process and we shall see that the limit theorems of the last section apply to $Y(t)$. It has been pointed out by Jagers (1967) that $\psi_i(x, \theta, t) = E(x^{X(t)}e^{-\theta Y(t)} \mid X(0) = i)$ $(\theta \geqq 0,\ 0 \leqq x \leqq 1)$ satisfies $\psi_i = (\psi_1)^i$ and

$$\psi_1(x, \theta, t) = e^{-(\theta+\nu)t}\left[x + \nu\int_0^t e^{(\theta+\nu)u}A(\psi_1(x, \theta, u))\,du\right]. \tag{24}$$

This integral equation implies that

$$\frac{\partial\psi_1}{\partial t} = a(\psi_1) - \theta\psi_1, \qquad \psi_1(x, \theta, 0) = x, \tag{25}$$

and since $(X(t), Y(t))$ is a bivariate Markov process we obtain the semi-group relation

$$\psi_1(x, \theta, t + \tau) = \psi_1(\psi_1(x, \theta, t), \theta, \tau). \tag{26}$$

Taking $\int_\tau^{t+\tau} Y(s)ds$ to be the functional $S_{\tau, t+\tau}$ of the last section, we see that (24) and (26) ensure the fulfiment of conditions (I)-(IV). An alternative approach, more in the spirit of Harris' (1963) definition of an M.B.P., may be taken by starting with (25) and the corresponding forward equation and deducing the existence of a suitable solution satisfying (26) and (24).

We now indicate the proof of the following analogue of Lemma 1.

Lemma 2. *Let* $\psi(\theta)$ *be the least positive solution of*

$$a(\psi) = \theta\psi. \tag{27}$$

Then for $0 \leqq x \leqq 1$, $\psi(x,\theta,t) \uparrow \psi(\theta)$ *(resp.* $\downarrow$*) as* $t \to \infty$ *if* $x \leqq \psi(\theta)$ *(resp.* $\geqq$*) uniformly for* $\theta \geqq 0$. *If* $a < 0$ *and* $A''(1-) < \infty$,

$$[\psi(\theta) - \psi(x,\theta,t)]e^{-tT(\theta)} \to \Gamma(x,\theta) \qquad (t \to \infty) \tag{28}$$

uniformly for $\theta \geqq 0$, *where*

$$T(\theta) = a'(\psi(\theta)) - \theta \tag{29}$$

and

$$\Gamma(x,\theta) = \begin{cases} \Gamma(0,\theta)\exp T(\theta) \displaystyle\int_0^x (a(y) - y\theta)^{-1} dy & (0 \leqq \psi(\theta) < x) \\ \Gamma(1,\theta)\exp T(\theta) \displaystyle\int_x^1 (a(y) - y\theta)^{-1} dy & (\psi(\theta) < x \leqq 1). \end{cases} \tag{30}$$

The proof requires the following extension of a result of Kingman (1963 b, Theorem 2).

Lemma 3. *Let* $f(\theta,t)$ *be defined on* $S \times (0,\infty)$ *and continuous in* t *for each* $\theta \in S$, *an infinite subset of the real line. If, for each* $h > 0$, $f(\theta, nh) \to 0$ $(n \to \infty)$ *uniformly for* $\theta \in S$, *then* $f(\theta,t) \to 0$ $(t \to \infty)$ *uniformly for* $\theta \in S$.

Proof. For each $\theta \in S$ and ε, $h > 0$ let $T(\theta,\varepsilon) = \sup\{t \mid |f(\theta,t)| \geqq \varepsilon\}$ and $T_h(\theta,\varepsilon) = \sup\{n \mid |f(\theta,nh)| \geqq \varepsilon\}$. The uniform convergence hypotheses imply that $T_h(\varepsilon) = \sup_{\theta\in S} T_h(\theta,\varepsilon) < \infty$. Suppose that $f(\theta,t)$ does not converge uniformly. Then $T(\theta,2\varepsilon)$ is not bounded on S and so there exists a sequence $\{\theta_n\}_{n=1}^{\infty}$ in S such that $T(\theta_n,2\varepsilon) \to \infty$ $(n \to \infty)$. Thus there is a sequence of points $\{t_n\}$ with $t_n \to \infty$ such that $|f(\theta_n,t_n)| > \varepsilon$. Continuity implies that there is a sequence of open intervals $\{I_n\}$ such that $t_n \in I_n$ and

$$|f(\theta_n,t)| > \varepsilon \text{ when } t \in I_n \qquad (n = 1,2,\cdots). \tag{31}$$

Now $G = \bigcup_n I_n$ is open and unbounded so there exists $h > 0$ such that $nh \in G$ for infinitely many n; see Kingman (1963 b, p. 595). Thus there exists an unbounded sequence $\{u_n\}$ such that $T_h(\varepsilon) < u_n \in G$ and $|f(\theta,u_n)| < \varepsilon$ for all $\theta \in S$ and n. This contradiction of (31) proves the Lemma.

Proof of Lemma 2. Taking $\tau = h$, $t = (n-1)h$; $h > 0$ in (26), the mean value theorem shows that $\{\psi_1(x,\theta,nh)\}$ is a monotone sequence. Thus $\psi_1(x,\theta,t) \to \phi(x,\theta)$, which satisfies

$$\phi(x,\theta) = \psi_1(\phi(x,\theta),\theta,t). \tag{32}$$

Differentiating this with respect to t and using (25) shows that ϕ satisfies (26). It is now possible to prove the first part of the Lemma for $t = nh$ $(h > 0,\ n \to \infty)$ by proceeding as in Pakes (1971 b, Lemma 2), and the general result then follows from Lemma 3.

Let $\Delta(\theta,t) = (\partial/\partial x)\psi_1|_{x=\psi(\theta)}$. Differentiating (24) with respect to x, setting $x = \psi(\theta)$ and using (32) yields an integral equation from which it follows that $\Delta(\theta,t) = \exp(tT(\theta))$. The proof of (28) for t of the form nh follows as in Pakes (1971 b, Lemma 3) and the general case again follows from Lemma 3.

Finally to obtain (30) note that (26) and (28) yield

$$\begin{aligned} \Gamma(x,\theta) &= \lim_{\tau\to\infty} e^{-(t+\tau)T(\theta)}[\psi(\theta) - \psi(x,\theta,t+\tau)] \\ &= e^{-tT(\theta)}\Gamma(\psi(x,\theta,t),\theta). \end{aligned}$$

Differentiating this relation with respect to t and using (25) yields

$$(a(\psi) - \theta\psi)(\partial/\partial x)\Gamma(x,\theta)\big|_{x=\psi} = T(\theta)e^{tT(\theta)}\Gamma(x,\theta) = T(\theta)\Gamma(\psi,\theta)$$

which is equivalent to the differential equation

$$(a(x) - \theta x)(\partial/\partial x)\Gamma(x,\theta) = T(\theta)\Gamma(x,\theta),$$

the solution of which is just (30).

Let $p_{ij}(\theta,t) = E(e^{-\theta Y(t)};\ X(t) = j\,|\,X(0) = i)$ $(i,j \in \mathscr{I})$ and $b = \nu A''(1-)$. With Lemma 2 in hand we can mimic the proof of Theorem 9 to obtain

Theorem 22. *Let* $a < 0$, $b < \infty$. *Then for* $i,j \in \mathscr{I}$

$$e^{-tT(\theta)}p_{ij}(\theta,t) \to i(\psi(\theta))^{i-1}\pi_j(\theta) \qquad (t \to \infty)$$

uniformly for $\theta \geqq 0$, *where* $\Sigma_{j\in\mathscr{I}}\pi_j(\theta)x^j = \Gamma(x,\theta) - \Gamma(0,\theta)$ *and both* $\Sigma_{j\in\mathscr{I}}\pi_j(\theta)$ *and* $\Sigma_{j\in\mathscr{I}}j\pi_j(\theta)$ *converge uniformly and are continuous for* $\theta \geqq 0$.

Thus the matrix $[p_{ij}(\theta,t);\ i,j \in \mathscr{I}]$ is uniformly $\lambda(\theta)$-positive, where $\lambda(\theta) = -T(\theta)$ and $e_i(\theta) = i(\psi(\theta))^{i-1}$. Thus if the initial distribution has a finite mean, Theorems 16 and 18 apply with $\mu = 1 - b/a$, which follows from (27) and (29). Similarly if $c = a'''(1-) < \infty$, Theorem 17 and 19 apply with

$$\sigma^2 = (c + b(1+\mu))/a^2.$$

Writing $p_{ij}(t) = p_{ij}(0,t)$ and letting T_j be the time to extinction when $X(0) = j$, we have $P(T_j \leqq t) = p_{j0}(t) = (p_{10}(t))^j$ whence, from (25),

$$f_j(t) = j(p_{10}(t))^{j-1}a(p_{10}(t)).$$

Thus Theorem 20 also applies with μ and σ^2 as above.

Denote the total number of deaths in $(0,t)$ by $D(t)$ and the total progeny up to time t by $N(t) = D(t) + X(t)$. Puri (1969) has obtained some results for $D(t)$ and $Y(t)$. It is clear that $D(t)$ and $N(t) - X(0)$ are functionals of the type considered in §5. Furthermore, it is not difficult to show that, for the M.B.P. with the present conditions, Theorems 16–20 apply, with

$$\mu = a + \nu(1 - b/a),\ \sigma^2 = -(\nu^2/a)[1 - 2b(1 + \nu/a) + b^2/\nu^2 - c/a].$$

Consider now a population in which objects reproduce as in an M.B.P. and which is also subject to an independent immigration component. New objects arrive according to a compound Poisson process with parameter Ω, $B(x)$ being the p.g.f. of the number of new arrivals at an immigration epoch. The population size, $Z(t)$, at time t is a Markov process with irreducible state space $\mathscr{S}$ which may be taken as a subset of the non-negative integers with $\{0\} \in \mathscr{S}$. This process is called a Markov branching process with immigration (M.B.I.) and it has been studied by Sevast'yanov (1957) and Pakes (1973a). Let $I(t) = \int_0^t Z(\tau)d\tau$,

$$p_{ij}(\theta,t) = E(e^{-\theta I(t)};\ Z(t) = j\,|\,Z(0) = i),\ \phi_i(x,\theta,t) = \sum_{j\in\mathscr{S}} p_{ij}(\theta,t)x^j$$

and $b(x) = -\Omega(1 - B(x))$. Then $I(t)$ is a functional of the type considered in § 5 and, from

$$\phi_i(x,\theta,t) = \psi_i(x,\theta,t)\exp\int_0^t b(\psi_1(x,\theta,u))du$$

it is easily seen that conditions (I)–(IV) are satisfied. Let $d = b'(1-)$ and $e = b''(1-)$. Combining the type of argument used in Pakes (1973 c, Theorem 1) and the proof of Lemma 2 gives the following result.

Theorem 23. *Let* $a < 0,\ b,\ d < \infty$. *Then*

$$e^{t\lambda(\theta)}p_{ij}(\theta,t) \to (\psi(\theta))^i\pi_j(\theta) \qquad (t \to \infty)$$

uniformly for $\theta \geqq 0$, *where* $\lambda(\theta) = -b(\psi(\theta))$,

$$\Lambda(x,\theta) = \sum_{j=0}^{\infty} \pi_j(\theta)x^j = \Lambda(0,\theta)\exp\int_0^x \frac{b(y)+\lambda(\theta)}{a(y)-\theta y}dy, \quad (0 \leqq x \leqq 1)$$

and the integrand is defined at $y = \psi(\theta)$ *so as to be continuous there. In addition,* $\sum_{j\in\mathcal{S}}\pi_j(\theta)$ *and* $\sum_{j\in\mathcal{S}} j\pi_j(\theta)$ *are uniformly convergent and continuous for* $\theta \geqq 0$.

Thus we see that Theorem 21 applies with $\mu = -d/a$ and $\sigma^2 = (2d - bd/a + e)/a^2$ where $e < \infty$. Similarly Theorem 21 applies to the total number of deaths by time t and to the total number of objects by time t with

$$\mu = -vd/a,\ \sigma^2 = [vd(1+2v/a) - v^2bd/a^2 + v^2e/a^2]/a.$$

REFERENCES

CHUNG, K. L. (1967). *Markov Chains with Stationary Transition Probabilities*, 2nd ed. Berlin: Springer.

FELLER, W. (1970). *An Introduction to Probability Theory and its Applications*, vol. 1, 3rd ed. New York: Wiley.

FELLER, W. (1971). *An Introduction to Probability Theory and its Applications*, vol. 2, 2nd ed. New York: Wiley.

FUKUSHIMA, M. & HITSUDA, M. (1967). On a class of Markov processes taking values on lines and the central limit theorem. *Nagoya Math. J.* **30**, 47–56.

HARRIS, T. E. (1963). *The Theory of Branching Processes*. Berlin: Springer.

IBRAGIMOV, I. A. & LINNIK, YU. V. (1971). *Independent and Stationary Sequences of Random Variables*. Groningen: Wolters-Noordhoff.

JAGERS, P. (1967). Integrals of branching processes. *Biometrika*, **54**, 263–71.

KINGMAN, J. F. C. (1961). A convexity property of positive matrices. *Quart. J. Math. Oxford* (2), **12**, 283–4.

KINGMAN, J. F. C. (1963 a). The exponential decay of Markov transition probabilities. *Proc. London Math. Soc.* (3) **13**, 337–58.

KINGMAN, J. F. C. (1963 b). Ergodic properties of continuous-time Markov processes and their discrete skeletons. *Proc. London Math. Soc.* (3) **13**, 593–604.

KOZAKIEWICZ, W. (1947). On the convergence of sequences of moment generating functions. *Ann. Math. Statist.* **18**, 61–9.

MATTHEWS, J. P. (1970). A central limit theorem for absorbing Markov chains. *Biometrika*, **57**, 129–39.

MATTHEWS, J. P. (1971). *A Study of Processes Associated with a Finite Markov Chain*. Unpublished Dissertation, Sheffield.

NAGAEV, S. V. (1957). Some limit theorems for stationary Markov chains. *Theory Prob. Appl.* **2**, 378–406.

PAKES, A. G. (1971 a). Branching processes, with immigration. *J. Appl. Prob.* **8**, 32–42.

PAKES, A. G. (1971b). Some limit theorems for the total progeny of a branching process. *Adv. Appl. Prob.* **3**, 176–92.

PAKES, A. G. (1973). On Markov branching processes with immigration. *Sankhyā A*. To appear.

PAKES, A. G. (1973 b). Conditional limit theorems for a left continuous random walk. *J. Appl. Prob.* **10**, 39–53.

PAKES, A. G. (1973 c). On dams with Markovian inputs. *J. Appl. Prob.* **10**. 317–29.

PAPANGELOU, F. (1968). A lemma on the Galton-Watson process and some of its consequences. *Proc. Amer. Math. Soc.* **22**, 1469–79.

PURI, P. S. (1969). Some limit theorems on branching processes and certain related processes. *Sankhyā A*, **31**, 57–74.

PURI, P. S. (1971). A method for studying the integral functionals of stochastic processes with applications: I. Markov chain case. *J. Appl. Prob.* **8**, 331–43.

REVESZ, P. (1968). *The Laws of Large Numbers.* New York: Academic Press.

ROCKAFELLAR, R. T. (1970) *Convex Analysis.* Princeton Univ. Press.

ROSENBLATT, M. (1971) *Markov Processes. Structure and Asymptotic Behaviour.* Berlin: Springer.

RUBIN, H. & VERE-JONES, D. (1968). Domains of attraction for the subcritical Galton-Watson process. *J. Appl. Prob.* **5**, 216–19.

SENETA, E. (1969). Functional equations and the Galton-Watson process. *Adv. Appl. Prob.* **1**, 1–42.

SENETA, E. & VERE-JONES, D. (1966). On quasi-stationary distributions in discrete-time Markov chains with a denumerable infinity of states. *J. Appl. Prob.* **3**, 403–34.

SERFLING, R. J. (1968). Contributions to central limit theory for dependent variables. *Ann. Math. Statist.* **39**, 1158–75.

SEVAST'YANOV, B. A. (1957). Limit theorems for branching processes of special form. *Theory Prob. Appl.* **2**, 321–31.

VERE-JONES, D. (1967). Ergodic properties of non-negative matrices — I. *Pacific J. Math.* **22**, 361–86.

VERE-JONES, D. (1969). Some limit theorems for evanescent processes. *Aust. J. Statist.* **11**, 67–78.

An identity for stopping times of a Markov process

J. W. PITMAN
University of Sheffield

0. Introduction

Suppose that $\{X_n, n = 0, 1, \cdots\}$ is a Markov process defined on a probability space $(\Omega, \mathscr{F}, P)$ with general state space $(S, \mathscr{S})$ and stationary transition probabilities. This paper is concerned with an identity which provides a useful expression for the expectation of certain random variables that are of interest when the Markov process $\{X_n\}$ is studied in conjunction with a stopping time of the process. This identity is stated below as (0.1) but it is first necessary to develop some notation.

Let $(S^\infty, \mathscr{S}^\infty)$ be the countable product space of all sequences $(x_0, x_1, \cdots)$ of elements of S, with the usual product σ-field. For each $n = 0, 1, \cdots$ the shifted process $\mathbf{X}_n$ is defined to be the mapping $\omega \to (X_n(\omega), X_{n+1}(\omega), \cdots)$ of Ω into S^∞. The shifted process $\mathbf{X}_n$ may be thought of as the original process $\mathbf{X}_0$ viewed from n on. Suppose that T is a stopping time of the process $\{X_n\}$. It is frequently possible to express random variables of interest in the form $\sum_{n=0}^{T-1} \Phi(\mathbf{X}_n)$ where Φ is a function defined on S^∞. The central result of this paper is the identity (0.1) below which gives a simple expression for the expectation of such random variables in terms of the pre-T occupation measure υ defined by

$$\upsilon(A) = E \sum_{n=0}^{T-1} I_A(X_n), \qquad A \in \mathscr{S},$$

where E denotes expectation with respect to P and I_A is the indicator function of the set A. ($\upsilon(A)$ is just the expected occupation time of the set A by $\{X_n\}$ before T.) The identity may be stated as follows: for any non-negative $\mathscr{S}^\infty$-measurable function Φ

$$E \sum_{n=0}^{T-1} \Phi(\mathbf{X}_n) = \int_S \upsilon(dx)\, E_x\Phi, \tag{0.1}$$

where $E_x\Phi$ is a suitably chosen version of the conditional expectation $E(\Phi(\mathbf{X}_0) \mid X_0 = x)$.

Suppose for example that T is the first passage time T^B to a set B in $\mathscr{S}$, and consider the corresponding first passage time function Φ^B defined on S^∞ by

$$\Phi^B(x_0, x_1, \cdots) = \inf\{n : n > 0,\ x_n \in B\},$$

so that $T^B = \Phi^B(\mathbf{X}_0)$. Now it is easy to see that $\Phi^B(\mathbf{X}_n)$ equals $T^B - n$ on the set $\{T^B > n\}$, and so

$$\sum_{n=0}^{T^B-1} \Phi^B(\mathbf{X}_n) = \sum_{n=0}^{T^B-1} (T^B - n) = \tfrac{1}{2}T^B(T^B + 1).$$

Thus the identity (0.1) yields

$$E\tfrac{1}{2}T^B(T^B + 1) = \int_S \upsilon^B(dx) m_{xB},$$

where υ^B denotes the pre-T^B occupation measure and m_{xB} is the expected first passage time to B, starting at x. Since ET^B is just $\upsilon^B(S)$ this identity rearranges to give a simple expression for the second moment of T^B:

$$E(T^B)^2 = \int_S \upsilon^B(dx)(2m_{xB} - 1).$$

In particular, when $\{X_n\}$ is a Markov chain with denumerable state space the integral on the right-hand side of (0.1) becomes simply a sum over all states, giving in this case

$$E(T^B)^2 = \sum_{x \in S} \upsilon^B(x)(2m_{xB} - 1),$$

where $\upsilon^B(x)$ is now the expected number of visits to the state x before hitting B.

The main identity (0.1) is proved by a straightforward argument in §2 after some preliminaries and development of notation in §1. Some examples are then provided which illustrate the use of the identity, and in a typical application the identity is used to prove a formula for factorial moments of first passage times in a recurrent Markov process. The reader is referred to Pitman (1972), for a detailed discussion of further applications in this vein including results for transient and absorbing Markov processes.

In §3 the main identity is used to establish some general relations between the distribution of a Markov process stopped at a random time T and the pre-T occupation measure. When applied to first passage times these results yield some interesting generalizations of known identities. In §4 the main identity is used to extend to Markov additive processes a well known identity due to Wald for the expectation of a randomly stopped sum of independent and identically distributed random variables. Finally, §5 is devoted to a brief discussion of some other identities for stopping times of a Markov process which can be established by arguments similar to those used to prove the main identity.

Most of the results of this paper have analogues for continuous parameter Markov processes, and discussion of these results will be taken up elsewhere.

1. NOTATION AND PRELIMINARIES

Let $(G, \mathscr{G})$ stand for any one of the measure spaces considered in this paper. A function g on G is supposed to be extended real valued unless otherwise specified, and g is said to be $\mathscr{G}$-measurable if it is measurable with respect to $\mathscr{G}$ and the usual Borel σ-field of the extended real line. The space of all non-negative $\mathscr{G}$-measurable functions is denoted by $\mathscr{G}_+$, and integrals and conditional expectations are supposed to be defined in the usual way for all functions in $\mathscr{G}_+$. Let g be a $\mathscr{G}$-measurable function. The positive and negative parts of g, g^+ and g^-, are defined as functions in $\mathscr{G}_+$ by $g^+ = \max(g,0)$, $g^- = -\min(g,0)$; for a measure γ on $\mathscr{G}$ the integral $\int g d\gamma$ is defined to equal $\int g^+ d\gamma - \int g^- d\gamma$ whenever this expression is not of the form $\infty - \infty$. Also, $\int g d\gamma$ is supposed to be defined in the obvious way for all functions g which are $\mathscr{G}$-measurable but only defined γ-almost everywhere on G.

Let $(S, \mathscr{S})$ be a measurable space.

A kernel on $(S, \mathscr{S})$ is a mapping $(x, A) \to K(x, A)$ of $S \times \mathscr{S}$ into the non-negative extended real line which has the following properties:

(i) for each x in S the mapping $A \to K(x, A)$ is a measure on $\mathscr{S}$,

(ii) for each A in $\mathscr{S}$ the mapping $x \to K(x, A)$ is $\mathscr{S}$-measurable.

A Markov kernel or transition probability function on $(S, \mathscr{S})$ is a kernel P such that $P(x, S) = 1$ for all x in S. A Markov kernel P will in future be written as $P(\cdot, \cdot)$ to avoid confusion with a probability measure P. Associated with a kernel K on $(S, \mathscr{S})$ are two operators $\boldsymbol{K}$, one defined on the space of all measures on $\mathscr{S}$ to act on the right of measures μ according to

$$(\mu \boldsymbol{K})(A) = \int_S \mu(dx)\, K(x, A), \qquad A \in \mathscr{S},$$

the other defined on the space $\mathscr{S}_+$ to act on the left of functions f according to

$$(\boldsymbol{K}f)(x) = \int_S K(x, dy) f(y), \qquad x \in S;$$

for an arbitrary $\mathscr{S}$-measurable function f, $\boldsymbol{K}f$ is supposed to be defined as above whenever the integral is defined for all points x in S. If K and L are two kernels on $(S, \mathscr{S})$ their composition KL is defined to be the kernel on $(S, \mathscr{S})$ which is the mapping $(x, A) \to \int_S K(x, dy)\, L(y, A)$. When the space S is denumerable, with $\mathscr{S}$ the power set of S, it is usual to identify a kernel K on $(S, \mathscr{S})$ with the matrix $(K(i, \{j\}))_{i,j \in S}$; operations with kernels then correspond in an obvious way to operations with their associated matrices.

Let N denote the set of non-negative integers. Unless otherwise specified n ranges over N. Throughout this paper it is supposed that $P(\cdot, \cdot)$ is a given Markov kernel on $(S, \mathscr{S})$, and that $\{X_n\}$ is a Markov process defined on a probability space $(\Omega, \mathscr{F}, P)$ with state space $(S, \mathscr{S})$ and stationary transition probability function $P(\cdot, \cdot)$: that is to say, for all A in $\mathscr{S}$ and n in N

$$P(X_{n+1} \in A \,|\, X_0, \cdots, X_n) = P(X_n, A) \text{ a.s..}$$

The sub-σ-field of $\mathscr{F}$ generated by the random variables $X_0, \cdots, X_n$ will be denoted by $\mathscr{F}_n$, and $\mathscr{F}_\infty$ is the sub-σ-field of $\mathscr{F}$ generated by the whole process $\{X_n\}$. When $\{\mathscr{G}_n\}$ is an increasing sequence of sub-σ-fields of $\mathscr{F}$ let us say that $\{X_n\}$ is Markov with respect to $\{\mathscr{G}_n\}$ if $\mathscr{F}_n \subseteq \mathscr{G}_n$ for all n in N, and for all A in $\mathscr{S}$ and n in N

$$P(X_{n+1} \in A \,|\, \mathscr{G}_n) = P(X_n, A) \qquad \text{a.s..}$$

We note that $\{X_n\}$ is Markov with respect to $\{\mathscr{F}_n\}$.

Let $(S^\infty, \mathscr{S}^\infty)$ be the countable product space of all sequences $\boldsymbol{x} = (x_0, x_1, \cdots)$ of elements of S, with the usual product σ-field. The Markov process $\{X_n\}$ may be considered in a natural way as the mapping $\boldsymbol{X}$ of Ω into S^∞ defined by $\boldsymbol{X}(\omega) = (X_0(\omega), X_1(\omega), \cdots)$, $\omega \in \Omega$. The mapping $\boldsymbol{X}$ is measurable and the distribution of $\boldsymbol{X}$ is determined by the initial distribution μ of X_0 and the kernel $P(\cdot, \cdot)$ of the Markov process. Indeed

$$P\{\boldsymbol{X} \in \Lambda\} = P_\mu(\Lambda), \qquad \Lambda \in \mathscr{S}^\infty, \tag{1.1}$$

where P_μ denotes the probability measure induced on $\mathscr{S}^\infty$ by μ and $P(\cdot, \cdot)$ in the following way: for $A_0, \cdots, A_n$ in $\mathscr{S}$ define

$$P_\mu\{\boldsymbol{x}: \boldsymbol{x} \in \mathscr{S}^\infty, x_0 \in A_0, \cdots, x_n \in A_n\} = \int_{A_0} \mu(dx_0) \int_{A_1} P(x_0, dx_1) \cdots \int_{A_n} P(x_{n-1}, dx_n);$$

this defines P_μ as a countably additive set function on the cylinder sets of $\mathscr{S}^\infty$, and the extension to $\mathscr{S}^\infty$ is unique (Doob, 1953, p. 613). For an arbitrary measure μ on $\mathscr{S}$ it is useful to continue to define a measure P_μ on $\mathscr{S}^\infty$ just as above, even though the probabilistic interpretation (1.1) is lost when μ is not a probability. For $\mathscr{S}^\infty$-measurable functions Φ let us write $E_\mu\Phi$ instead of $\int \Phi dP_\mu$, and when μ is the probability measure concentrated at the point x in S let us write P_x for P_μ and E_x for E_μ. The function $x \to E_x\Phi$ is $\mathscr{S}$-measurable on the domain of definition of $E_x\Phi$, and if μ is a measure on $\mathscr{S}$ it follows from Fubini's theorem that

$$E_\mu\Phi = \int_S \mu(dx)\, E_x\Phi \tag{1.2}$$

whenever $E_\mu\Phi$ is defined. It should be noted from (1.1) that if $\{X_n\}$ has initial distribution μ then $E\Phi(X) = E_\mu\Phi$ for all $\mathscr{S}^\infty$-measurable Φ for which either quantity is defined. Also, for Φ in $\mathscr{S}^\infty_+$ the function $x \to E_x\Phi$ is a well defined and convenient version of the conditional expectation $E(\Phi(X) \mid X_0 = x)$. Let X_n denote the measurable mapping $\omega \to (X_n(\omega), X_{n+1}(\omega), \cdots)$ of Ω into S^∞. The Markov property of the process $\{X_n\}$ can be succinctly expressed as follows: $\{X_n\}$ is Markov with respect to $\{\mathscr{G}_n\}$ if and only if for each n in N

$$E(\Phi(X_n) \mid \mathscr{G}_n) = E_{X_n}\Phi \text{ a.s. for all } \Phi \text{ in } \mathscr{S}^\infty_+. \tag{1.3}$$

A random variable defined on $(\Omega, \mathscr{F}, P)$ with values in $N \cup \{\infty\}$ will be referred to as a *random time.* Let $\{\mathscr{G}_n\}$ be an increasing sequence of sub-σ-fields of $\mathscr{F}$. A random time T is said to be a *stopping time of* $\{\mathscr{G}_n\}$ if the event $\{T > n\}$ is in $\mathscr{G}_n$ for all n in N. A stopping time of $\{\mathscr{F}_n\}$ is also said to be a stopping time of $\{X_n\}$. More generally, let us make the following definition (Chow & Robbins, 1971; Pitman & Speed (to appear) and Siegmund, 1967).

Definition 1.4 A random time T is said to be a *randomized stopping time of* $\{X_n\}$ if for each n in N

$$P(T > n \mid \mathscr{F}_\infty) = P(T > n \mid \mathscr{F}_n) \qquad \text{a.s..}$$

A randomized stopping time of $\{X_n\}$ can be thought of as being generated in the following way: an observer watches the evolution of the process $\{X_n\}$ as time n increases, until a random time T that he "stops"; if at time k he has not yet stopped, the observer notes the new value X_k and then decides whether to stop at time k or to continue to observe the process according to the outcome of some random experiment. The random time T is a randomized stopping time of $\{X_n\}$ if the reasonable assumption is made that for each k the random experiment at time k and the as yet unobserved future $\{X_n, k < n < \infty\}$ are conditionally independent given the observed past $\{X_n, 0 \leq n \leq k\}$. For example, the observer might decide whether or not to stop at time k by tossing a coin chosen with bias some measurable function of $(X_0, \cdots, X_k)$. The random time T is a stopping time of $\{X_n\}$ if for each k the decision at time k is made deterministically (and measurably) according to $(X_0, \cdots, X_k)$. This interpretation of randomized stopping times and the important Proposition 1.5 below are both consequences of a fundamental property of conditional independence (II, T51 in Meyer, 1966). For a more detailed discussion of these matters see Pitman & Speed (to appear).

For a random time T let us denote by $\mathscr{F}_n^T$ the smallest sub-σ-field of $\mathscr{F}$ containing $\mathscr{F}_n$ and the events $\{T = 0\}, \cdots, \{T = n\}$. Obviously T is a stopping time of $\{\mathscr{F}_n^T\}$.

Proposition 1.5. A random time T is a randomized stopping time of the Markov process $\{X_n\}$ if and only if $\{X_n\}$ is Markov with respect to $\{\mathscr{F}_n^T\}$.

Proof. See the remarks above.

Let ξ_n denote the nth co-ordinate mapping $(x_0, x_1, \cdots) \to x_n$ of S^∞ into S. If μ is a probability measure on $\mathscr{S}$ the process $\{\xi_n\}$ defined on $(S^\infty, \mathscr{S}^\infty, P_\mu)$ is Markov with initial distribution μ and kernel $P(\cdot,\cdot)$, and by virtue of (1.1) the distribution of the general Markov process $\{X_n\}$ with initial distribution μ and kernel $P(\cdot,\cdot)$ can be studied entirely through the co-ordinate process $\{\xi_n\}$ on $(S^\infty, \mathscr{S}^\infty, P_\mu)$. This approach does not always suit our purpose though; we are about to consider the general Markov process $\{X_n\}$ in conjunction with a randomized stopping time T of $\{X_n\}$, and working with the co-ordinate process $\{\xi_n\}$ would be tantamount to assuming that T was a stopping time of $\{X_n\}$, with consequent loss of generality. However, when considering stopping times of a Markov process it will often be convenient to work with the co-ordinate process, and in this case no loss of generality is incurred by doing so.

2. The main identity

Suppose that $\{X_n\}$ is a Markov process defined on the probability space $(\Omega, \mathscr{F}, P)$ with state space $(S, \mathscr{S})$ and transition probability function $P(\cdot,\cdot)$. Let T be a randomized stopping time of $\{X_n\}$.

Definition 2.1. The pre-T occupation measure is the set function υ from $\mathscr{S}$ into the non-negative extended real line defined by

$$\upsilon(A) = E \sum_{n=0}^{T-1} I_A(X_n), \qquad A \in \mathscr{S}.$$

Thus $\upsilon(A)$ is the expected occupation time of the set A by $\{X_n\}$ before T. It is easily seen that

$$\upsilon(A) = \sum_{n=0}^{\infty} P\{X_n \in A,\ T > n\}, \qquad A \in \mathscr{S}, \tag{2.2}$$

which makes it obvious that υ is in fact a measure on $\mathscr{S}$. Note that $\upsilon(S) = ET$. It is important to observe that for all functions f in $\mathscr{S}_+$

$$E \sum_{n=0}^{T-1} f(X_n) = \int_S f\, d\upsilon. \tag{2.3}$$

Indeed, (2.3) holds by definition for indicator functions f, and hence for simple functions; the monotone convergence theorem then extends (2.3) to all functions f in $\mathscr{S}_+$. It may be observed that Definition 2.1 makes sense for any random time T even if $\{X_n\}$ is not Markov, and (2.3) holds in this generality.

As an example consider the stopping time T^B of $\{X_n\}$ which is the first passage time to the set B in $\mathscr{S}$, $T^B = \inf\{n : n > 0,\ X_n \in B\}$, (inf of the empty set is taken to be ∞). The quantities appearing in the sum (2.2) are then simply the taboo probabilities

$$P\{X_n \in A,\ T^B > n\} = P\{X_n \in A;\ X_k \notin B,\ 0 < k \leq n\}$$

(cf. Chung, 1967, §9). The identity (2.3) generalizes Theorem 14.5 of Chung (1967), and (1.4) of Orey (1959).

We now come to the main result.

Theorem 2.4. Suppose that T is a randomized stopping time of the Markov process $\{X_n\}$, and let υ denote the pre-T occupation measure. Then for all $\mathscr{S}^\infty$-measurable functions Φ for which $E_\upsilon\Phi$ is defined

$$E \sum_{n=0}^{T-1} \Phi(X_n) = E_\upsilon\Phi = \int_S \upsilon(dx)\, E_x\Phi.$$

Remark. The equality of the central and right-hand expressions is nothing more than (1.2).

Proof. Suppose first that $\Phi \in \mathscr{S}^\infty_+$. Consider

$$\begin{aligned} E \sum_{n=0}^{T-1} \Phi(X_n) &= E \sum_{n=0}^{\infty} \Phi(X_n) I_{\{T>n\}} \\ &= \sum_{n=0}^{\infty} \int_{\{T>n\}} \Phi(X_n)dP \qquad \text{by Fubini's theorem.} \end{aligned} \tag{2.5}$$

As for Proposition 1.5 let $\mathscr{F}_n^T$ be the sub-σ-field of $\mathscr{F}$ generated by the random variables $X_0, \cdots, X_n$ and the events $\{T = 0\}, \cdots, \{T = n\}$. Now

$$\begin{aligned} \int_{\{T>n\}} \Phi(X_n)dP &= \int_{\{T>n\}} E(\Phi(X_n) \,|\, \mathscr{F}_n^T)dP \\ &\qquad\qquad \text{since } \{T > n\} \in \mathscr{F}_n^T \\ &= \int_{\{T>n\}} (E_{X_n}\Phi)dP \end{aligned}$$

by Proposition 1.5 and the Markov property (1.3). But substituting this in (2.5) and using Fubini's theorem again gives

$$\begin{aligned} E \sum_{n=0}^{T-1} \Phi(X_n) &= E \sum_{n=0}^{\infty} (E_{X_n}\Phi) I_{\{T>n\}} \\ &= E \sum_{n=0}^{T-1} E_{X_n}\Phi = \int_S \upsilon(dx)E_x\Phi \end{aligned}$$

by (2.3). This proves the theorem for Φ in $\mathscr{S}^\infty_+$. The result for general Φ follows after writing $\Phi = \Phi^+ - \Phi^-$ and applying the theorem to Φ^+ and Φ^- separately.

It may be noted that $E_\upsilon\Phi$ is defined and the identity of the theorem holds whenever either (i) Φ is non-negative, or (ii) Φ is bounded and ET is finite.

It is interesting to consider the theorem in a slightly different way: a measure Υ can be defined on $\mathscr{S}^\infty$ by

$$\Upsilon(\Lambda) = E \sum_{n=0}^{T-1} I_\Lambda(X_n), \qquad \Lambda \in \mathscr{S}^\infty,$$

so that $\Upsilon(\Lambda)$ is the expected number of times that the sequence of shifted processes $\{X_n\}$ lies in the set Λ before time T; just as for (2.3) one has for Φ in $\mathscr{S}^\infty_+$

$$E \sum_{n=0}^{T-1} \Phi(X_n) = \int_S \Phi d\Upsilon,$$

and thus the theorem simply states that $\Upsilon = P_\upsilon$.

To illustrate the theorem let us consider a simple example. For A, B in $\mathscr{S}$ let N_{AB} denote the random variable on Ω which is the number of one-step transitions from A to B before T:

$$N_{AB} = \sum_{n=0}^{T-1} \Phi_{AB}(X_n)$$

where $\Phi_{AB} = I_A(\xi_0)I_B(\xi_1)$ is in $\mathscr{S}_+^\infty$. Immediately from the theorem one has a simple formula for the expectation of N_{AB}:

$$EN_{AB} = E_v\Phi_{AB} = \int_A v(dx)P(x, B).$$

Suppose for example that $\{X_n\}$ is a recurrent Markov chain with irreducible state space S and that $X_0 = i$ a.s. for some i in S. As is well known the expected number of visits of $\{X_n\}$ to a state j before the time of the first return to i is simply $\pi(j)/\pi(i)$, where π is an invariant measure for the transition function $P(\cdot, \cdot)$. According to the above formula the expected number of $j \to k$ transitions before the first return to i is just $\pi(j)P(j, \{k\})/\pi(i)$, $j, k \in S$.

More generally, let h be a non-negative $\mathscr{S} \times \mathscr{S}$-measurable function on $S \times S$. Taking $\Phi = h(\xi_0, \xi_1)$ in the theorem gives

$$E \sum_{n=0}^{T-1} h(X_n, X_{n+1}) = E_v h(\xi_0, \xi_1) = \int_S v(dx) \int_S P(x, dy)\, h(x, y). \tag{2.6}$$

For a more interesting application of the theorem let us now turn to the consideration of return times in a recurrent Markov process. Let ϕ be a σ-finite measure on $\mathscr{S}$, and suppose that the kernel $P(\cdot\,, \cdot)$ induces a ϕ-recurrent Markov process. That is to say, for all A in $\mathscr{S}$ such that $\phi(A) > 0$ and all x in S

$$P_x\{\xi_n \in A \text{ for some } n > 0\} = 1$$

where $\{\xi_n\}$ is the co-ordinate process on $(S^\infty, \mathscr{S}^\infty)$. For such a kernel $P(\cdot, \cdot)$ there exists a σ-finite invariant measure π on $\mathscr{S}$, $(\pi \mathbf{P} = \pi)$, and π is unique up to a constant factor (Orey, 1971, Theorem 7.2). Let us take a particular set B in $\mathscr{S}$ with $0 < \pi(B) < \infty$ and assume for convenience that π has been chosen so that $\pi(B) = 1$. Let $\pi \mid B$ denote the probability on $\mathscr{S}$ which is the restriction of π to B.

Consider now the Markov process $\{\xi_n\}$ with initial distribution $\pi \mid B$ defined on $(S^\infty, \mathscr{S}^\infty, P_{\pi|B})$. Let τ^B be the time of the first return of $\{\xi_n\}$ to the set B, $\tau^B = \inf\{n: n > 0, \xi_n \in B\}$. The random time τ^B is a stopping time of $\{\xi_n\}$, and it is well known that the pre-τ^B occupation measure for $\{\xi_n\}$ is simply π. It then follows from the theorem that for any $\mathscr{S}^\infty$-measurable function Φ for which $E_\pi\Phi$ is defined

$$E_{\pi|B} \sum_{n=0}^{\tau^B - 1} \Phi(\xi_n) = E_\pi \Phi, \tag{2.7}$$

a formula which may be viewed as a generalization of a result due to Orey (1959, Lemma 6).

As an application of (2.7) let us consider an identity for factorial moments of the return time τ^B. For n, r in N, $r \geqq 1$, let

$$(n)_r = n(n-1)\cdots(n-r+1).$$

By taking Φ to be $(\tau^B)_r$ in (2.7) we can prove the following identity: for all $r \geqq 1$,

$$E_{\pi|B}(\tau^B)_{r+1} = (r+1)E_{\pi|B^c}(\tau^B)_r. \tag{2.8}$$

To see this let us observe that

$$\tau^B(\xi_n) = \tau^B - n \text{ on the set } \{\tau^B > n\}, \tag{2.9}$$

and use the fact that

$$(n+1)_{r+1} - (n)_{r+1} = (r+1)(n)_r. \tag{2.10}$$

Now

$$\begin{aligned}(r+1)\sum_{n=0}^{\tau^B-1}(\tau^B(\xi_n))_r &= (r+1)\sum_{n=0}^{\tau^B-1}(\tau^B-n)_r \text{ by (2.9)}\\ &= (\tau^B+1)_{r+1} = (\tau^B)_{r+1} + (r+1)(\tau^B)_r\end{aligned}$$

by first telescoping the sum with (2.10) and then using (2.10) again. Thus, integrating with respect to $P_{\pi|B}$,

$$\begin{aligned}E_{\pi|B}(\tau^B)_{r+1} + (r+1)E_{\pi|B}(\tau^B)_r &= (r+1)E_{\pi|B}\sum_{n=0}^{\tau^B-1}(\tau^B(\xi_n))_r\\ &= (r+1)E_\pi(\tau^B)_r\end{aligned}$$

by applying (2.7) with $\Phi = (\tau^B)_r$, and (2.8) follows at once.

When the state space is denumerable the set B can be taken to be a single point $\{i\}$, and in this case (2.8) gives rise to the following identity for moments of first passage times in an irreducible recurrent Markov chain with invariant measure π: for $r \geqq 1$ and all states i,

$$E_i(\tau^{\{i\}})_{r+1} = \frac{(r+1)}{\pi(i)}\sum_{j\neq i}\pi(j)E_j(\tau^{\{i\}})_r.$$

This identity has been proved in a completely different way by Lamperti (1960).

The identity (2.8) is but one of a large number of identities for first passage times which can be established by applying the theorem after using the property (2.9) of first passage times τ^B to express a random variable of interest as $\Sigma_{n=0}^{\tau^B-1}\Phi(\xi_n)$ for some $\mathscr{S}^\infty$-measurable function Φ related to τ^B. This idea is systematically exploited in Pitman (1972).

3. THE PRE-T OCCUPATION MEASURE AND THE DISTRIBUTION OF THE PROCESS STOPPED AT T

Suppose we are given a Markov process $\{X_n\}$ defined on $(\Omega, \mathscr{F}, P)$ with state space $(S, \mathscr{S})$, stationary transition probability function $P(\cdot,\cdot)$ and initial distribution μ^0. In this section Theorem 2.4 is used to establish some identities which relate the pre-T occupation measure and the distribution of the process stopped at T for randomized stopping times T of the process $\{X_n\}$.

Let us use the following convention: when Φ is an $\mathscr{S}^\infty$-measurable function and T is a random time on Ω the random variable $\Phi(X_T)$ is defined everywhere on Ω by

$$\begin{aligned}\Phi(X_T) &= \Phi(X_n) \quad \text{on } \{T=n\}, \qquad n\in N\\ &= 0 \qquad\quad \text{on } \{T=\infty\}.\end{aligned}$$

For an $\mathscr{S}$-measurable function f the random variable $f(X_T)$ is defined on Ω in a similar way, consistent with writing $f(X_T) = (f\circ\xi_0)(X_T)$.

For a random time T let μ^T denote the sub-probability on $\mathscr{S}$ which is the distribution of the partially defined random variable X_T, the process stopped at T:

$$\mu^T(A) = EI_A(X_T) = \sum_{n=0}^{\infty}P\{T=n,\, X_n\in A\}, \qquad A\in\mathscr{S}.$$

The pre-T occupation measure will be denoted by υ^T:

$$\upsilon^T(A) = E \sum_{n=0}^{T-1} I_A(X_n) = \sum_{n=0}^{\infty} P\{T > n, X_n \in A\}, \qquad A \in \mathscr{S}.$$

The measures μ^T and υ^T are very similar in character and it may be observed that Theorem 2.4 is analogous to the following integrated form of the strong Markov property: for any randomized stopping time T of $\{X_n\}$

$$E\Phi(X_T) = E_{\mu^T}\Phi \tag{3.1}$$

for all $\mathscr{S}^\infty$-measurable functions Φ for which either quantity is defined.

Proposition 3.2. If T is a randomized stopping time of $\{X_n\}$ the measures υ^T and μ^T are related by

$$\upsilon^T + \mu^T = \mu^0 + \upsilon^T \boldsymbol{P}.$$

Remark. Letting ψ denote the measure appearing on either side of this identity, $\psi(A)$ is the expected occupation time of the set A by $\{X_n\}$ up to and including time T.

Proof. Let T be a randomized stopping time of $\{X_n\}$. It is evident that

$$\sum_{n=0}^{T-1} I_A(X_n) + I_A(X_T) = I_A(X_0) + \sum_{n=0}^{T-1} I_A(X_{n+1}).$$

Taking expectations in this equation and applying Theorem 2.4 with $\Phi = I_A(\xi_1)$ to the second term on the right-hand side gives

$$\upsilon^T(A) + \mu^T(A) = \mu^0(A) + \int_S \upsilon^T(dx)\, P(x, A), \qquad A \in \mathscr{S},$$

which is just Proposition 3.2.

Let A be a set in $\mathscr{S}$. It is an immediate consequence of Proposition 3.2 that for any randomized stopping time T of $\{X_n\}$ one has $\upsilon^T(A) = \infty$ if and only if $\upsilon^T \boldsymbol{P}(A) = \infty$, and if $\upsilon^T(A) < \infty$ then

$$\mu^T(A) = (\mu^0 + \upsilon^T \boldsymbol{P} - \upsilon^T)(A). \tag{3.3}$$

Thus, if $ET (= \upsilon^T(S))$ is finite the distribution of X_T is uniquely determined by the pre-T occupation measure; indeed μ^T is determined by υ^T through (3.3) whenever υ^T is σ-finite. Another consequence of Proposition 3.2 is that, if T is such that the distribution of X_T is the same as the initial distribution, then the pre-T occupation measure is invariant for the kernel $P(\cdot, \cdot)$, $\upsilon^T \boldsymbol{P} = \upsilon^T$. Conversely, if υ^T is σ-finite and invariant then $\mu^T = \mu^0$. A well known example of this is afforded by return times of a recurrent Markov process (see the end of the last section).

Suppose now that we are given a randomized stopping time T of $\{X_n\}$ such that the distribution of X_T is μ^T. To what extent is the pre-T occupation measure influenced by μ^T? Assuming for the moment that ET is finite, it follows from Proposition 3.2 that the pre-T occupation measure must be a solution of the equation

$$\upsilon(\boldsymbol{I} - \boldsymbol{P}) = \lambda, \tag{3.4}$$

where $\boldsymbol{I}$ denotes the identity operator, $\lambda = \mu^0 - \mu^T$ is a known signed measure, and υ is an unknown measure. The question of what measures υ satisfy this equation is related to the potential theory associated with the kernel $P(\cdot,\cdot)$ (see Kemeny, Snell & Knapp, 1966). However, many of the results which emerge from the potential theory can be proved more simply and intuitively by working with Proposition 3.5 below which provides a useful complement to the identity of Proposition 3.2.

For Proposition 3.5 let B be a set in $\mathscr{S}$, and define the first passage function τ^B in $\mathscr{S}_+^\infty$ by $\tau^B = \inf\{n: n > 0, \xi_n \in B\}$. For A in $\mathscr{S}$ define χ_A^B in $\mathscr{S}_+^\infty$ to be the pre-τ^B occupation time of A by the co-ordinate process $\{\xi_n\}$:

$$\chi_A^B = \sum_{n=0}^{\tau^B - 1} I_A(\xi_n),$$

and for x in S let

$$U^B(x, A) = E_x \chi_A^B.$$

That is to say, $U^B(x,\cdot)$ is the pre-τ^B occupation measure for the Markov process $\{\xi_n\}$ with initial distribution concentrated at x and kernel $P(\cdot,\cdot)$. We note that U^B is a kernel on $(S, \mathscr{S})$, and if T^B denotes the first passage time to B of the process $\{X_n\}$, $T^B = \tau^B(X)$, then the pre-T^B occupation measure for $\{X_n\}$ is simply $\mu^0 U^B$. For a measure λ on $\mathscr{S}$ and a set C in $\mathscr{S}$ denote the restriction of λ to C by $\lambda | C$: $(\lambda | C)(A) = \lambda(C \cap A)$, $A \in \mathscr{S}$.

Proposition 3.5. For any randomized stopping time T of $\{X_n\}$

$$\upsilon^T + (\mu^T | B^c) U^B = (\mu^0 | B^c) U^B + (\upsilon^T | B) U^B.$$

Remark. Letting ψ denote the measure which appears on either side of this identity, $\psi(A)$ is the expected number of times that the process $\{X_n\}$ visits the set A before the time of the first visit to B at or after time T.

Proof. Let us show that for all A in $\mathscr{S}$

$$\sum_{n=0}^{T-1} I_A(X_n) + I_{B^c}(X_T)\chi_A^B(X_T) = I_{B^c}(X_0)\chi_A^B(X_0) + \sum_{n=0}^{T-1} I_B(X_n)\chi_A^B(X_n). \tag{3.6}$$

Taking expectations in this equation, applying the strong Markov property (3.1) and Theorem 2.4 (with Φ taken to be $I_B(\xi_0)\chi_A^B$) then shows that for all A in $\mathscr{S}$

$$\upsilon^T(A) + \int_{B^c} \mu^T(dx) U^B(x, A) = \int_{B^c} \mu^0(dx) U^B(x, A) + \int_B \upsilon^T(dx) U^B(x, A),$$

which is what is required.

To see the truth of (3.6) let $0 \leqq T_1 < T_2 < \cdots$ be the random variables defined on Ω as the successive times that the process $\{X_n\}$ visits the set B: $T_1 = \inf\{n: n \in N, X_n \in B\}$, $T_k = \inf\{n: n \in N, n > T_{k-1}, X_n \in B\}$, $k \geqq 2$, and let $T_\infty = \infty$. Let M be the pre-T occupation time of B by $\{X_n\}$. The left-hand side of (3.6) is just the number of visits to the set A before T_{M+1}, the time of the first visit to B at or after time T. But so too is the right-hand side of (3.6): the first term is the number of visits to A before the first visit to B (at time T_1), and the factor $I_B(X_n)$ in the sum ensures that the only terms which can be non-negative are the T_1th, $\cdots$, T_Mth, the kth of these being just the number of visits to A between the times T_k and T_{k+1}, $1 \leqq k \leqq M$.

Let us now assume that the state space S is denumerable with $\mathscr{S}$ the power set of S, so that the process $\{X_n\}$ is a Markov chain with transition matrix $\boldsymbol{P} = (P(i, \{j\}))_{i,j \in S}$ and initial distribution μ^0. Suppose that T is a randomized stopping time of the Markov chain $\{X_n\}$

such that X_T has distribution μ^T. Let us see what can be deduced from Proposition 3.5 concerning the pre-T occupation measure. Our first observation is that the pre-T occupation measure of transient points is completely determined by μ^T. Let $\upsilon_i(j) = U^{\phi}(i, \{j\})$, where ϕ denotes the empty set. (The matrix $(\upsilon_i(j))_{i,j \in S}$ is just $\Sigma_{n=0}^{\infty} \boldsymbol{P}^n$.)

Proposition 3.7. If j is a transient state of the Markov chain then

$$\upsilon^T(j) = \sum_i [\mu^0(i) - \mu^T(i)]\upsilon_i(j).$$

Proof. Take $B = \phi$ in Proposition 3.5 and evaluate the measures at the set $\{j\}$. The second term on the right-hand side is obviously zero, and the identity can be rearranged to yield Proposition 3.7, since j transient implies $\upsilon_i(j) \leqq \upsilon_j(j) < \infty$ for all i in S.

A state which is not transient must belong to some recurrent class, and for this case we have Proposition 3.8 below. For i, j, k in S let

$$\upsilon_i^k(j) = U^{\{k\}}(i, \{j\}),$$

the expected number of visits to j before the first visit to k for initial distribution concentrated at i.

Proposition 3.8. Let k be a state in a recurrent communicating class C. Then for all states j in C

$$\upsilon^T(j) = \upsilon^T(k)\upsilon_k^k(j) + \sum_{i \neq k} [\mu^0(i) - \mu^T(i)]\upsilon_i^k(j).$$

Proof. Take $B = \{k\}$ in Proposition 3.5 and evaluate the measures at the set $\{j\}$. The identity rearranges to give Proposition 3.8 since j and k belonging to the same recurrent class implies that for all i in S

$$\upsilon_i^k(j) \leqq \upsilon_j^k(j) < \infty.$$

In an irreducible Markov chain, Proposition 3.8 can be proved in a different way when ET is finite, as a consequence of (3.4) and some results in the potential theory of recurrent Markov chains (Theorem 9.15(1) and Corollaries 9.17 and 9.54 of Kemeny, Snell & Knapp, 1966).

According to Proposition 3.8 the pre-T occupation is determined on the whole of a recurrent class C by its value at any one point k in C and the distribution of X_T. It is an interesting consequence of Propositions 3.7 and 3.8 that for any randomized stopping time T of a Markov chain the finiteness of $\upsilon^T(j)$ is a class property: that is to say, in a communicating class C either $\upsilon^T(j)$ is finite for all j in C or $\upsilon^T(j)$ is infinite for all j in C. Obviously this latter can only occur in a recurrent class.

If the Markov chain is irreducible and positive recurrent it is possible to assert much more than Proposition 3.8. Let m_{ij} be the mean first passage time from i to j,

$$m_{ij} = E_i\tau^{\{j\}} = \upsilon_i^j(S).$$

In this case m_{ij} is finite for all i and j in S and we have the following result:

Proposition 3.9. If T is a randomized stopping time of an irreducible positive recurrent Markov chain and $ET < \infty$ then, for all states j in S

$$\upsilon^T(j) = \frac{1}{m_{jj}} \left\{ET + \sum_k \sum_{i \neq j} [\mu^T(i) - \mu^0(i)]\upsilon_i^j(k)\right\},$$

and if furthermore $\Sigma_i \mu_0(i) m_{ij} < \infty$ for some (and hence every) state j in S then also

$$\upsilon_T(j) = \frac{1}{m_{jj}} \left[ET + \sum_{i \neq j} [\mu^T(i) - \mu^0(i)] m_{ij} \right]. \tag{3.10}$$

Proof. To obtain the first result simply interchange the roles of j and k in Proposition 3.8 and then sum over k. When $\Sigma_i \mu_0(i) m_{ij} < \infty$ the double series is absolutely convergent and the second formula results on changing the order of summation.

Thus, for a positive recurrent irreducible Markov chain $\{X_n\}$ the pre-T occupation measure for any randomized stopping time of $\{X_n\}$ with finite expectation is completely determined by ET and the distribution of X_T. It is further clear from (3.10) that when μ^0 is such that $\Sigma_{i \neq j} \mu^0(i) m_{ij} < \infty$ for some (and hence all) j in S, $\upsilon^T(j) = \infty$ for all j in S if and only if $ET = \infty$. This is not the case when $\Sigma_{i \neq j} \mu^0(i) m_{ij} = \infty$ for j in S: the stopping time $T^{\{j\}}$ (which is a.s. finite) has infinite expectation but the pre-$T^{\{j\}}$ occupation measure of $\{k\}$ is not greater than $\upsilon_k^j(k)$ which is finite for all k in S.

To conclude this section let us consider some consequences of Proposition 3.5 for first passage times. It is convenient to work with a slight modification of Proposition 3.5, considering occupation times after time zero and up to and including a random time T. Define the measure $\bar{\upsilon}^T$ on $\mathscr{S}$ by

$$\bar{\upsilon}^T(A) = E \sum_{n=1}^{T} I_A(X_n) = \sum_{n=1}^{\infty} P\{T \geqq n, X_n \in A\}, \qquad A \in \mathscr{S}.$$

From Proposition 3.2 and its proof it is clear that the measures $\bar{\upsilon}^T$ and υ^T are related by

$$\bar{\upsilon}^T = \upsilon^T \boldsymbol{P} = \upsilon^T + \mu^T - \mu^0. \tag{3.11}$$

Define the kernel $\bar{U}^B$ on $(S, \mathscr{S})$ by

$$\bar{U}^B(x, A) = E_x \sum_{n=1}^{\tau B} I_A(\xi_n).$$

It is easy to see that Proposition 3.5 becomes

$$\bar{\upsilon}^T + \mu^T \bar{\boldsymbol{U}}^B = \mu^0 \bar{\boldsymbol{U}}^B + (\bar{\upsilon}^T \mid B) \bar{U}^B. \tag{3.12}$$

Letting ψ denote the measure which appears on either side of (3.12), $\psi(A)$ is the expected number of times that the process $\{X_n\}$ visits the set A after time zero and up to and including the time of the first visit to B after time T.

When K is a kernel on $(S, \mathscr{S})$ and B is a set in $\mathscr{S}$ let us denote by $K \mid B$ the kernel defined by

$$(K \mid B)(x, A) = K(x, B \cap A), \quad x \in S, \ A \in \mathscr{S}.$$

Thus for x in S, $(K \mid B)(x, \cdot)$ is the restriction of $K(x, \cdot)$ to B.

Proposition 3.13. For all sets B and C in $\mathscr{S}$,

$$\bar{U}^C + (\bar{U}^C \mid C) \bar{U}^B = \bar{U}^B + (\bar{U}^C \mid B) \bar{U}^B.$$

Remark. Letting K denote the kernel which appears on either side of this identity, $K(x, A)$ is E_x {the number of times that $\{\xi_n\}$ visits A after time zero and up to and including the time of the first visit to B after the time of the first visit to C}.

Proof. Apply (3.12) to the stopping time τ^C of $\{\xi_n\}$, using the fact that the P_x distribution of the process stopped at τ^C is just $\bar{U}^C(x,\cdot)$ restricted to C.

Proposition 3.13 yields Corollary 3.14 below for Markov chains. For i and j in a denumerable state space S, let $\bar{\upsilon}_i^j$ be the measure $\bar{U}^{\{j\}}(i,\cdot)$, ($\bar{\upsilon}_i^j(k)$ is the ${}_jp_{ik}^*$ of Chung, 1967, §I.9). By virtue of (3.11) the measures $\bar{\upsilon}_i^j$ and υ_i^j determine each other by $\bar{\upsilon}_i^j = \upsilon_i^j \boldsymbol{P}$ and

$$\upsilon_i^j(k) = \bar{\upsilon}_i^j(k) + \delta_{ik} - \delta_{jk}\bar{\upsilon}_i^j(j), \qquad k \in S$$

where $\delta_{lm} = 1$ if $l = m$, 0 otherwise. Note that $\bar{\upsilon}_i^j(j) = P_i\{\tau^{\{j\}} < \infty\}$, which is always one in an irreducible recurrent chain.

Corollary 3.14. For all states i, j and k of a Markov chain,

$$\bar{\upsilon}_i^k + \bar{\upsilon}_i^k(k)\bar{\upsilon}_k^j = \bar{\upsilon}_i^j + \bar{\upsilon}_i^k(j)\bar{\upsilon}^j.$$

Proof. Take $B = \{j\}$ and $C = \{k\}$ in Proposition 3.13.

For a different proof see Kemeny & Snell (1961), Lemma 6. Recall that m_{ij} denotes the mean first passage time $E_i\tau^{\{j\}}$. Evaluating the identity of Corollary 3.14 at the whole set S yields the following identity after observing that m_{ik} is infinite if $\bar{\upsilon}_i^k(k)$ is not equal to one:

$$m_{ik} + m_{kj} = m_{ij} + \bar{\upsilon}_i^k(j)m_{jj}.$$

When m_{ij} and m_{jj} are both finite (as is the case when i and j belong to the same positive recurrent class) this identity rearranges to give the following simple expression for $\bar{\upsilon}_i^k(j)$ in terms of mean first passage times:

$$\bar{\upsilon}_i^k(j) = (m_{ik} + m_{kj} - m_{ij}) \,/\, m_{jj}.$$

This is a formula due to Chung (1967, I.11).

4. An identity for Markov additive processes

Let $\{X_n\}$ be a Markov process defined on a probability space $(\Omega, \mathscr{F}, P)$ with state space $(S, \mathscr{S})$. Suppose that there is also defined on $(\Omega, \mathscr{F}, P)$ a sequence $\{Y_n\}$ of real-valued random variables such that for each $k \geqq 1$ the distribution of Y_k depends on the Markov process $\{X_n\}$ only through the states X_{k-1} and X_k, and the sequence $\{Y_n\}$ is conditionally independent given the Markov process $\{X_n\}$. That is to say, letting $(R, \mathscr{R})$ denote the real line with the Borel σ-field, the bivariate process $\{(X_n, Y_n)\}$ is a Markov process with state space $(S \times R, \mathscr{S} \times \mathscr{R})$ and transition probability function $\hat{P}(\cdot,\cdot)$, for which there exists a function $Q(\cdot,\cdot)$ on $S \times (\mathscr{S} \times \mathscr{R})$ such that for all x in S and D in $\mathscr{S} \times \mathscr{R}$

$$\hat{P}((x, y), D) = Q(x, D), \qquad y \in R. \tag{4.1}$$

Define random variables S_n on Ω by $S_0 = 0$

$$S_n = \sum_{k=1}^{n} Y_k, \qquad n \geqq 1.$$

The bivariate process $\{(X_n, S_n)\}$ will be referred to as a *Markov additive process* (see Arjas & Speed, 1972). When the summands Y_n are a.s. positive the Markov additive process $\{(X_n, S_n)\}$ is closely associated with a Markov renewal process and a semi-Markov process (Pyke, 1961, and Çinlar, 1969).

If the Markov process $\{X_n\}$ has but one state, the random variables $Y_1, Y_2, \cdots$ are necessarily independent and identically distributed, and $\{S_n\}$ is simply a random walk on the line. Proposition 4.3 below provides an extension to general Markov additive processes of the following well known identity for this case due to Wald (1944): if T is a stopping time for a sequence $\{Y_n\}_{n \geqq 1}$ of independent and identically distributed random variables, and

$$S_n = \sum_{k=1}^{n} Y_k, \qquad n \geqq 1, \qquad \text{then}$$

$$ES_T = (EY_1)(ET) \tag{4.2}$$

whenever EY_1 is defined and $ET < \infty$.

For ease of notation it will be assumed (without loss of generality) that the Markov additive process $\{(X_n, S_n)\}$ is in fact derived from the co-ordinate process $\{(X_n, Y_n)\}$ defined on $((S \times R)^\infty, (\mathscr{S} \times \mathscr{R})^\infty, P)$, where P is a probability measure $\hat{P}_\lambda$ on $(\mathscr{S} \times \mathscr{R})^\infty$ induced in the usual way by some initial distribution λ on $\mathscr{S} \times \mathscr{R}$, and a transition probability function $\hat{P}(\cdot, \cdot)$ on $(S \times R, \mathscr{S} \times \mathscr{R})$ which is supposed to satisfy (4.1).

Let δ_0 denote the probability measure on $\mathscr{R}$ which is degenerate at zero. A measure μ on $\mathscr{S}$ will be identified with the measure $\mu \times \delta_0$ on $\mathscr{S} \times \mathscr{R}$; thus E_μ will be written for $E_{\mu \times \delta_0}$, and E_x for $E_{(x,0)}$.

Note that S_0 is defined to be zero (not Y_0), and for $1 \leqq n \leqq \infty$, $S_n = \sum_{k=1}^{n} Y_k$, where the random variable S_∞ is only defined on the set where the series converges.

Proposition 4.3. Let T be a stopping time of the Markov additive process $\{(X_n, S_n)\}$ and let υ denote the pre-T occupation measure for $\{X_n\}$. If $E_\upsilon Y_1$ is defined then

$$ES_T = E_\upsilon Y_1.$$

Proof. First observe that

$$\sum_{n=1}^{T} Y_n^+ = \sum_{n=0}^{T-1} Y_1^+(X_n, Y_n).$$

Now, a stopping time of $\{(X_n, S_n)\}$ is also a stopping time of $\{(X_n, Y_n)\}$, and hence letting $\hat{\upsilon}$ denote the pre-T occupation measure for $\{(X_n, Y_n)\}$, one has from Theorem 2.4 that

$$\begin{aligned} E \sum_{n=1}^{T} Y_n^+ &= E_{\hat{\upsilon}} Y_1^+ \\ &= E_\upsilon Y_1^+, \end{aligned} \tag{4.4}$$

by Fubini's theorem, using the facts that $\upsilon(A) = \hat{\upsilon}(A \times R)$, $A \in \mathscr{S}$, and that for x in S $E_{(x,y)} Y_1^+ = E_x Y_1^+$, $y \in R$. Similarly

$$E \sum_{n=1}^{T} Y_n^- = E_\upsilon Y_1^-. \tag{4.5}$$

Now $E_\upsilon Y_1$ is assumed to be defined, so that at least one of $E_\upsilon Y_1^+$ and $E_\upsilon Y_1^-$ is finite, and thus by (4.4) and (4.5) at least one of the random variables $\Sigma_{n=1}^T Y_n^+$ and $\Sigma_{n=1}^T Y_n^-$ is finite a.s.. Hence S_T is defined and equal to $(\Sigma_{n=1}^T Y_n^+) - (\Sigma_{n=1}^T Y_n^-)$ a.s., and

$$ES_T = E\sum_{n=1}^{T} Y_n^+ - E\sum_{n=1}^{T} Y_n^- = E_\upsilon Y_1^+ - E_\upsilon Y_1^- = E_\upsilon Y_1.$$

When $ET < \infty$ the identity of Proposition 4.3 can be made to appear very similar to the Wald identity (4.2). Let α be the pre-T occupation measure υ normalized to be a probability on $\mathscr{S}$, $\alpha(A) = \upsilon(A)/\upsilon(S)$, $A \in \mathscr{S}$, where of course $\upsilon(S) = ET < \infty$; one has immediately from Proposition 4.3

$$ES_T = (E_\alpha Y_1)(ET) \tag{4.6}$$

whenever $E_\alpha Y_1$ is defined, and (4.6) reduces to (4.2) when the Markov chain $\{X_n\}$ has only a single state. Any stopping time for $\{(X_n, S_n)\}$ is a randomized stopping time for $\{X_n\}$, and it follows from (3.3) that α is an invariant probability measure for the Markov process $\{X_n\}$ if the distribution of X_T is the same as the distribution of X_0. The identity (4.6) has been proved for this case by Arjas (see Arjas & Speed, 1972, 2.32). For a concrete example see Pyke & Schaufele (1964), Lemma 4.1 or Hunter (1969), Theorem 2.9.

A particularly simple Markov additive process arises when for some $\mathscr{S} \times \mathscr{S}$-measurable function h, $Y_n = h(X_{n-1}, X_n)$ a.s.. For this case Proposition 4.5 reduces to the identity (2.6) given in §2.

5. Related identities

In this final section it is observed that thc arguments used to prove the main identity can easily be modified to yield some other identities of interest. Throughout this section $\{X_n\}$ is a Markov process defined on a probability space $(\Omega, \mathscr{F}, P)$ with state space $(S, \mathscr{S})$ and transition probability function $P(\cdot,\cdot)$, and T is a randomized stopping time of $\{X_n\}$. The notation developed for this framework in §1 will be used without comment.

Proposition 5.1. Let $\{\Phi_n\}$ be a sequence of functions in $\mathscr{S}_+^\infty$. Then

$$E\sum_{n=0}^{T-1} \Phi_n(X_n) = E\sum_{n=0}^{T-1} E_{X_n}\Phi_n.$$

Proof. The identity is proved by an obvious adaptation of the proof of Theorem 2.4.

It is clear that the identity will hold for an arbitrary sequence $\{\Phi_n\}$ of $\mathscr{S}^\infty$-measurable functions, provided that the identity holds finitely with Φ_n replaced by $|\Phi_n|$. If $\{a_n\}$ is a sequence of non-negative real numbers one may define a pre-T occupation measure v, weighted by $\{a_n\}$, through

$$v(A) = E\sum_{n=0}^{T-1} a_n I_A(X_n) = \sum_{n=0}^{\infty} a_n P\{T > n, X_n \in A\}, \qquad A \in \mathscr{S}.$$

Just as for (2.3),

$$E\sum_{n=0}^{T-1} a_n f(X_n) = \int_S f\,dv, \tag{5.2}$$

and hence, from Proposition 5.1,

$$E \sum_{n=0}^{T-1} a_n \Phi(X_n) = E_\nu \Phi = \int_S \nu(dx) E_x \Phi \tag{5.3}$$

for all $\mathscr{S}^\infty$-measurable functions Φ for which $E_\nu \Phi$ is defined. In particular for $0 \leqq t < \infty$ let υ_t denote the pre-T occupation measure weighted by $\{t^n\}$:

$$\upsilon_t(A) = E \sum_{n=0}^{T-1} t^n I_A(X_n). \tag{5.4}$$

Then from (5.3),

$$E \sum_{n=0}^{T-1} t^n \Phi(X_n) = E_{\upsilon_t} \Phi = \int_S \upsilon_t(dx) E_x \Phi \tag{5.5}$$

for all $\mathscr{S}^\infty$-measurable functions Φ for which $E_{\upsilon_t}\Phi$ is defined. An important consequence of this identity is Proposition 5.6 below. For Proposition 5.6, let us define t^∞ to be zero, and let $\boldsymbol{I}$ denote the identity operator.

Proposition 5.6. If $0 \leqq t < 1$ and f is a bounded real-valued $\mathscr{S}$-measurable function,

$$Ef(X_0) - Et^T f(X_T) = \int_S (\boldsymbol{I} - t\boldsymbol{P}) f \, d\upsilon_t \tag{5.7}$$

$$= E \sum_{n=0}^{T-1} [(\boldsymbol{I} - t\boldsymbol{P}) f](X_n). \tag{5.8}$$

Proof. Define Φ on S^∞ by $\Phi = f(\xi_0) - tf(\xi_1)$. Then

$$\sum_{n=0}^{T-1} t^n \Phi(X_n) = \sum_{n=0}^{T-1} [t^n f(X_n) - t^{n+1} f(X_{n+1})]$$

$$= f(X_0) - t^T f(X_T),$$

since $0 \leqq t < 1$ and f is bounded. But $E_x \Phi = (\boldsymbol{I} - t\boldsymbol{P}) f(x)$ and $E_{\upsilon_t}\Phi$ is defined (and finite) because $E_x |\Phi|$ is bounded on S and υ_t is a finite measure on $\mathscr{S}$. Thus (5.7) follows from (5.5), and then (5.8) from (5.2).

In connection with Proposition 5.6 let us make the following observations:

(i) It is easy to see that the identities are equally valid for complex t with $|t| < 1$ and bounded complex valued $\mathscr{S}$-measurable functions f, with υ_t defined by (5.4) as a complex measure on $\mathscr{S}$.

(ii) The identities also hold for an arbitrary real-valued $\mathscr{S}$-measurable function f and arbitrary positive real t, provided that the expectations on the left-hand side are defined and finite, and $E \sum_{n=0}^{T-1} t^n |f(X_n)|$ $(= \int |f| d\upsilon_t)$ is finite. This can be proved like Proposition 3.2, using (5.5).

(iii) Proposition 5.6 could be reformulated as an identity for measures like Proposition 3.2, which would then be the special case $t = 1$.

Proposition 5.6 is proved by Arjas & Speed (1972), as a corollary of their fundamental identity for Markov additive processes. It may be noted that their identity can easily be established by viewing a Markov additive process as a bivariate Markov process and applying the results of this section.

The reader is referred to Kemperman (1961), for a variety of applications of Proposition 5.6 (Kemperman's Theorem 9.1 is a version of Proposition 5.6 for a special class of randomized stopping times.). For applications to Markov additive processes and semi-Markov processes see Arjas (1972), Arjas (1972a) and Arjas & Speed (1972).

Acknowledgement: I am grateful to Dr. T.P. Speed for many valuable comments.

References

ARJAS, E. (1972). On a fundamental identity in the theory of semi-Markov processes. *Adv. Appl. Prob.* **4,** 258–70.

ARJAS, E. (1972a). On the use of a fundamental identity in the theory of semi-Markov queues. *Adv. Appl. Prob.* **4,** 271–84.

ARJAS, E. & SPEED, T. P. (1972). Topics in Markov additive processes. Helsinki Univ. Tech., Inst. Math., Research Report A24.

ARJAS, E. & SPEED, T. P. (1972a). Symmetric Wiener-Hopf factorisations in Markov additive processes. Helsinki Univ. Tech., Inst. Math., Research Report A25.

CHOW, Y. S., ROBBINS, H. & SIEGMUND, D. (1971). *Great Expectations: The Theory of Optimal Stopping.* Boston: Houghton Mifflin Co.

CHUNG, K. L. (1967). *Markov Chains with Stationary Transition Probabilities.* 2nd Ed. Berlin: Springer-Verlag.

ÇINLAR, E. (1969). Markov renewal theory. *Adv. Appl. Prob.* **1,** 123–87.

DOOB, J. L. (1953). *Stochastic Processes.* New York: Wiley.

HUNTER, J. J. (1969). On the moments of Markov renewal processes. *Adv. Appl. Prob.* **1,** 188–210.

KEMENY, J. G. &. SNELL, J. L. (1961). Potentials for denumerable Markov chains. *J. Math. Anal. & Appl.* **3,** 196–260.

KEMENY, J. G., SNELL, J. L. & KNAPP, A. W. (1966). *Denumerable Markov Chains.* Princeton: Van Nostrand.

KEMPERMAN, J. H. B. (1961). *The Passage Problem for a Stationary Markov Chain.* Chicago: University of Chicago Press.

LAMPERTI, J. (1960). The first-passage moments and the invariant measure of a Markov chain. *Ann. Math. Statist.* **31,** 515–7.

MEYER, P. A. (1966). *Probability and Potentials.* Waltham, Mass.: Blaisdell.

OREY, S. (1959). Recurrent Markov chains. *Pacific J. Math.* **9,** 805–27.

OREY, S. (1971). *Limit Theorems for Markov Chain Transition Probabilities.* London: Van Nostrand.

PITMAN, J. W. (1972). Some identities for stopping times of a Markov chain. Univ. Sheffield, Dept. Prob. & Stat., Research Report 122.

PITMAN, J. W. & SPEED, T. P. (to appear). A note on random times. Univ. Sheffield, Dept. Prob. & Stat. Research Report.

PYKE, R. (1961). Markov renewal processes; definitions and preliminary properties. *Ann. Math. Statist.* **32** 1231–42.

PYKE, R. & SCHAUFELE, R. A. (1964). Limit theorems for Markov renewal processes. *Ann. Math. Statist.* **35,** 1746–64.

SIEGMUND, D. (1967). Some problems in the theory of optimal stopping rules. *Ann. Math. Statist.* **38,** 1627–40.

WALD, A. (1944). On cumulative sums of random variables. *Ann. Math. Statist.* **15,** 283–96.

Models of Topographic Variation

M. L. THORNETT
University of Western Australia

SUMMARY

The paper is concerned with developing models of topographic variation, i.e., models which might help in the analysis of experimental measurements associated with regions or "plots" in n-dimensional Euclidean space R^n. The view is taken that the measurements of interest may be regarded as observations on random variables indexed according to the bounded Borel subsets of R^n, these random variables having finite variances and satisfying certain stationarity conditions. By introducing appropriate notions of convergence, spectral representations are derived which are analogous to those already familiar from the theory of continuous stationary random processes. Fairfield Smith's well-known empirical law of agricultural variation is the prototype for a certain class of the models discussed.

1. INTRODUCTION

The (conceptual) starting point of many agricultural experiments is a rectangular region consisting of r rows and c columns of contiguous plots of dimensions a, b laid out as follows:

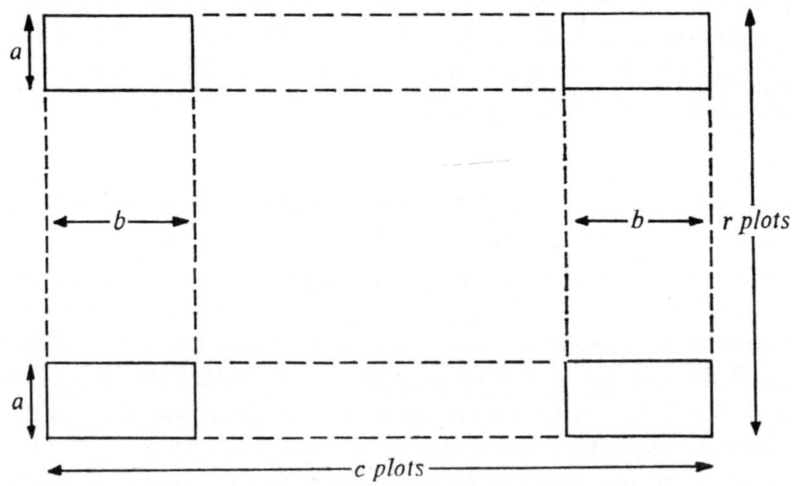

The nature of the data collected is, apart from the various treatments applied to the plots, a set of responses for each plot. If w_{ij} denotes one such response measurement for the plot in the ith row and jth column, then a basic statistic calculated from such data is

$$s^2\{a,b;r,c\} = \left\{\sum_{i=1}^{r}\sum_{j=1}^{c} w_{ij}^2 - \left(\sum_{i=1}^{r}\sum_{j=1}^{c} w_{ij}\right)^2 / (rc)\right\} / \{ab(rc-1)\}. \tag{1.1}$$

Assuming the w_{ij} to be observations on uncorrelated random variables W_{ij} having a common expectation μab and a common variance $\sigma^2 ab$, then $E(S^2\{a,b;r,c\})$ has the constant value σ^2 independent of the values taken by the four parameters a, b, r and c.

However, it is only rarely that experiments or uniformity trials support an assumed model of this kind. Typically, $s^2\{a,b;r,c\}$ is larger when the plot area ab is larger, and there is often the suggestion that long thin plots are less variable than square plots of the same area. A more fundamental observation is that neighbouring plots tend to yield more similarly than those further apart (cf. Matérn, 1960, §6.12).

It is for such reasons that the devices of randomization on the one hand, and of incomplete block designs on the other, loom so large in agricultural experimentation. Evidently, if independence of responses could be assumed, there would be no (statistical) problems about plot size and shape, relative location of plots and of treatments; in short, no problems of experimental design.

1.1 *Uniformity trials.* The have been many investigations of the nature of the dependence of the quantity $s^2\{a,b;r,c\}$ on its parameters, the experimental investigations usually taking the form of "uniformity trials". In such trials the whole of a certain large area is sown to the same crop or pasture, this area, at harvest time, being subdivided into a large number of small contiguous rectangular plots, the separate yields of which are obtained and recorded. Then, as a basis for inferring how $s^2\{a,b;r,c\}$ varies with the values of a, b, r, and c, the usual *ad hoc* procedure is to calculate this quantity for as many combinations of the parameters as the observations allow, or as are deemed adequate to accomplish this purpose.

1.2 *Fairfield Smith's law.* Possibly the most well-known and extensive of trials of this nature is that reported by H. Fairfield Smith (1938), the paper being noteworthy also for the large bibliography referring to other such trials. In this paper the conclusion is reached that, for the trial in question, and for many of the other trials, the dependence of $s^2\{a,b;r,c\}$ on plot shape is negligible, and that the empirical law based on

$$\mathrm{Var}(W_{ij}) = \sigma^2(ab)^\gamma \tag{1.2}$$

with $1 \leqq \gamma < 2$, is in reasonable accord with the data. One notes that $\gamma = 1$ corresponds to the situation of uncorrelated plot yields, while $\gamma = 2$ (formally) may be equated with complete dependence between plot yields, in which case of course, the quantities $s^2\{a,b;r,c\}$ would all be zero.

Fairfield Smith's law has proved so acceptable that many uniformity trials have subsequently been carried out for the express purpose of determining the parameters σ^2 and γ, the idea being that, by invoking various economic considerations, the optimum plot and block sizes for experiments on a given crop can thereby be deduced.

The variation in the plot yields of an agricultural experiment may be considered as an example in two dimensions of "topographic variation" (cf. Whittle, 1962), this being an all-embracing term for the variation in some arbitrary response measurement from region to region in n-dimensional Euclidean space R^n. This paper examines ways in which models might be developed to account for various instances of topographic variation, attempting in particular to give some theoretical foundation for Fairfield Smith's law.

2. Preliminaries

Apart from notation specifically developed for our purposes, we shall employ notation and terminology frequently if not commonly found in the literature. For example, we follow the convention of indicating a random variable by including somewhere in its representation a capital Latin letter, which, when replaced by the corresponding small Latin letter then represents an observation on the random variable in question. Generally however, such commonly used notation will be used without comment. At this stage it suffices to note the following:

2.1 *Notation.* The letters x, y, and u, with or without superscripts, will usually denote points in R^n, the particular coordinates of such points being denoted by the same letter subscripted, e.g., $x = (x_1, \cdots, x_n)$. Where relevant, the symbols representing points in R^n will be handled as though they were $n \times 1$ matrices. We put

$$u \cdot x = \sum_{j=1}^{n} u_j x_j \text{ and } |x| = \sqrt{x \cdot x}.$$

The remaining small Latin letters, when not associated with observations on random variables, will be employed as complex, real, or integer-valued variables, whichever is indicated by the context or convention. In particular, r will always denote a real number satisfying $1 \leqq r \leqq \infty$, p will denote a real number satisfying $1 \leqq p \leqq 2$, while by p' we shall understand $p/(p-1)$.

For any measure space $(S, \mathscr{E}, \mu)$, $\mathscr{C}[(S, \mathscr{E}, \mu)]$ will denote the set of equivalence classes of measurable complex-valued functions on S, while $L^r(S, \mu)$ will have the usual meaning. In the case of a probability space $(S, \mathscr{E}, P)$, the members of $\mathscr{C}[(S, \mathscr{E}, P)]$ will be identified with the distinct complex-valued random variables that may be defined on this space.

Small Greek letters will be used mostly to denote the members of $\mathscr{C}[(R^n, \mathscr{B}^n, \lambda^n)]$, where $\mathscr{B}^n$ is the class of Borel subsets of R^n and λ^n denotes n-dimensional Lebesgue measure. In particular

$$\chi_A(x) = \begin{matrix} 1 & \text{if } x \in A \\ 0 & \text{if } x \notin A \end{matrix}$$

will denote the indicator function of $A \in \mathscr{B}^n$. We put

$$\mathscr{B}^n_b = \text{the class of bounded Borel subsets of } R^n,$$

$$L^r_n = L^r(R^n, \lambda^n),$$

$$\tilde{\phi}(u) = \int \exp(i\, u \cdot x)\phi(x)dx \text{ for } \phi \in L^1_n,$$

$$\tau_y \phi(x) = \phi(x - y),$$

$$\phi^*(x) = \overline{\phi(-x)},$$

$$\phi * \psi(x) = \int \phi(y)\psi(x-y)dy, \text{ while}$$

$$\|\phi\|_r = \left[\int |\phi(x)|^r dx\right]^{1/r},$$

shall denote the norm on L_n^r. In addition to the subspaces L_n^r of $\mathscr{C}[(R^n, \mathscr{B}^n, \lambda^n)]$ we shall need to consider

L_n = the vector subspace consisting of those functions of the form $\sum_{j=1}^k c_j \chi_{A_j}(x)$ where c_j is complex and $A_j \in \mathscr{B}_b^n$, $j = 1, \cdots, k$; and

K_n = the vector subspace consisting of those functions which are infinitely differentiable and have compact supports.

Finally, the expectation and variance of any complex random variable X will be denoted by $E(X)$ and $\mathrm{Var}(X)$ respectively, and for any two complex random variables X and Y we shall use the inner product notation (X, Y) in place of $E(X\bar{Y})$. We note that

$$\mathrm{Cov}(X, Y) = (X, Y) - E(X)\overline{E(Y)}.$$

3. Model-building condiserations

In setting up a model for topographic variation in R^n our object, briefly stated, is to assign random variables to regions of R^n which have expectations and covariances compatible with certain invariance and convergence requirements.

Proceeding one step at a time, the first property to be embodied in the model is that the response measurement for any given plot in R^n must be the sum of the corresponding measurements for any finite collection of disjoint subplots whose union is the given plot. Thus we are interested in the existence of a probability space $(S, \mathscr{E}, P)$ and a mapping $X: \mathscr{B}_b^n \to L^2(S, P)$ such that

$$X(A \cup B) = X(A) + X(B) \tag{3.1}$$

whenever $A, B \in \mathscr{B}_b^n$ are disjoint.

3.1 *Linearity*. The above model is equivalently represented by a mapping $X: L_n \to L^2(S, P)$ which, for all $\phi, \psi \in L_n$ and complex constants c, d, satisfies the linearity condition

$$X(c\phi + d\psi) = cX(\phi) + dX(\psi), \tag{3.2}$$

this equivalence simply being a consequence of identifying $X(A)$ with $X(\chi_A)$ for all $A \in \mathscr{B}_b^n$. Moreover, in identifying the domain of X with L_n in this way we have gained the mathematical convenience of being able to deal with a family of random variables (the range of X) which is closed with respect to forming finite linear combinations.

3.2 *Translation-invariance*. For all $y \in R^n$ and $\phi_1, \cdots, \phi_k \in L_n$ we demand that $X(\phi_1), \cdots, X(\phi_k)$ have the same joint distribution as $X(\tau_y\phi_1), \cdots, X(\tau_y\phi_k)$.

In an agricultural context the above would mean that the yields of a given collection of plots would have the same joint distribution regardless of how these plots were translated relative to a given frame of reference. We could of course ask that the joint distribution of the yields of the plots be completely independent of orientation, i.e., invariant under rotations as well as translations. Such "isotropic" models will indeed be considered later. However, it is clear that there are many agricultural situations in which, due to the usual cultivating and planting techniques say, one would concede translation-invariance relative to a naturally imposed frame of reference, but would be unwilling to accept the assumption of isotropy.

3.3 *Continuity*. A further consideration in respect of any model we entertain is how the random variables $X(\phi)$ and $X(\psi)$ are to be related when the functions ϕ, $\psi \in L_n$ approach one another in some sense. This suggests the imposition of some kind of continuity condition on our mapping $X: L_n \to L^2(S, P)$.

We may therefore sum up our model-building problem as follows: For some suitable topological vector subspace, N_n say, of $\mathscr{C}[(R^n, \mathscr{B}^n, \lambda^n)]$ such that $N_n \supset L_n$, find a probability space $(S, \mathscr{E}, P)$, and look amongst the continuous translation-invariant linear mappings $X: N_n \to L^2(S, P)$ for prospective models of topographic variation in R^n. The fundamental problem raised by this paper is how to choose N_n in the most natural and/or convenient way.

3.4 *Implications for the functionals* E_X *and* C_X. Let us now examine the properties of the functionals E_X and C_X (or E and C when there is no risk of confusion) defined by

$$E_X(\phi) = E(X(\phi)) \text{ and } C_X(\phi, \psi) = (X(\phi), X(\psi)) \tag{3.3}$$

when the mapping $X: N_n \to L^2(S, P)$ satisfies the above linearity, invariance, and continuity conditions.

Firstly, E is a continuous translation-invariant linear functional on N_n. That is,

(E.1) if the functions $\phi_1, \phi_2, \cdots$, converge to zero in N_n then $E(\phi_m) \to 0$;

(E.2) $E(\tau_y \phi) = E(\phi)$ for all $\phi \in N_n$, $y \in R^n$; and

(E.3) $E(c\phi + d\psi) = cE(\phi) + dE(\psi)$ for all complex c, d.

As for the functional C, it has the following properties:

(C.1) If $\phi_1, \phi_2, \cdots$, converges to zero in N_n then $C(\phi_m, \phi_m) \to 0$;

(C.2) C is positive-definite, i.e., $0 < C(\phi, \phi) < \infty$ for all $\phi \in N_n$;

(C.3) C is translation-invariant, i.e., $C(\tau_y \phi, \tau_y \psi) = C(\phi, \psi)$ for all $\phi, \psi \in N_n$ and all $y \in R^n$;

(C.4) C is Hermitian, i.e., $C(\phi, \psi) = \overline{C(\psi, \phi)}$;

(C.5) for fixed ψ, C is linear in ϕ. Indeed, in view of (C.1), (C.2), and the Schwarz inequality, C is a continuous linear functional in ϕ for fixed ψ. Also, from (C.4), $\bar{C}$ is a continuous linear functional in ψ for fixed ϕ.

Thus, C may be referred to as a continuous positive-definite translation-invariant Hermitian bilinear functional. Henceforth the symbols E and C, possibly subscripted, will always stand for functionals having the above properties with respect to some topological vector space $N_n \subset \mathscr{C}[(R^n, \mathscr{B}^n, \lambda^n)]$. For the present we do not necessarily assume that $N_n \supset L_n$.

3.5 *Existence.* A question which may now be asked is whether, given some such N_n, there exist non-trivial continuous translation-invariant linear mappings from N_n into $L^2(S, P)$ for some underlying probability space $(S, \mathscr{E}, P)$. This can be answered in the affirmative by the Daniell-Kolmogorov theorem, provided we can exhibit the functionals E and C; since (cf. Gel'fand & Vilenkin, 1964, p. 252), all we need do is assign a complex normal distribution to $X(\phi_1), \cdots, X(\phi_k)$ for every selection $\phi_1, \cdots, \phi_k \in N_n$, the expectation and covariance matrices being determined by E and C in the obvious way. Moreover, this theorem shows that S and $\mathscr{E}$ may always be determined in terms of N_n, and we shall henceforth assume that this is the case, denoting them by R^{2N_n} and $\mathscr{B}^{2N_n}$ respectively (cf. Loève, 1963, p. 92). The factor 2 has been inserted since it is a complex rather than real random variable that is being associated with each $\phi \in N_n$.

3.6 *The collection* $G[N_n]$. For any N_n we shall use $G[N_n]$ to denote the collection of all mappings $X: N_n \to L^2(R^{2N_n}, P)$ for some probability measure P on $\mathscr{B}^{2N_n}$ such that the corresponding E_X and C_X have the properties (E.1) to (C.5). Thus we are henceforth following the usual practice of studying processes which are translation-invariant only as far as first and second moments are concerned—wide-sense translation-invariance—and are therefore not necessarily strictly translation-invariant as originally demanded in Section 3.2.

4. CONSIDERATIONS INVOLVED IN THE CHOICE N_n.

Perhaps the most significant step in the study of collections such as $G[N_n]$ has been the theory of "generalized random processes" as developed by Itô (1953), Yaglom (1957), and Gel'fand & Vilenkin (1964), in which N_n is taken as K_n (endowed with the Schwartz topology). We summarize the main results of this theory as follows:

4.1 *Theorem.* For every $X \in G[K_n]$ there exists (i) a Lebesgue-Stieltjes measure F on $\mathscr{B}^n$ satisfying

$$\int (1 + |u|^2)^{-k} F(du) < \infty \tag{4.1}$$

for some non-negative integer k; (ii) a complex constant c satisfying

$$|c|^2 \leqq F(\{0\}); \tag{4.2}$$

and (iii), a completely additive random set function Z on $\mathscr{B}_b^n$ satisfying

$$E(Z(A)) = c\chi_A(0), \; (Z(A), \overline{Z(B)}) = F(A \cap B) \tag{4.3}$$

for all $A, B \in \mathscr{B}_b^n$ such that

$$X(\phi) = \int \tilde{\phi}(u) Z(du), \tag{4.4}$$

$$E_X(\phi) = c\tilde{\phi}(0), \text{ and} \tag{4.5}$$

$$C_X(\phi, \psi) = \int \tilde{\phi}(u)\overline{\tilde{\psi}(u)} F(du) \tag{4.6}$$

for all $\phi, \psi \in K_n$.

The measure F is referred to as the spectral measure of the "process" X, while (4.4), (4.5), and (4.6) are called the spectral representations of X, E_X, and C_X respectively. Evidently, it changes nothing to let k be any real number, either negative or non-negative, but as it turns out, only the range $0 \leqq k < \infty$ is of interest. We shall use $k(F)$ to denote the greatest lower bound of all $k \geqq 0$ such that (4.1) is true.

4.2 *Properties required of* N_n. For our purposes however, the space K_n is inadequate since it does not contain L_n as a subspace; and while it is true (see Gel'fand & Shilov, 1964, p. 142) that there exist functions in K_n "arbitrarily close" to any given $\phi \in L_n$, the nature of this approximation is not enough to guarantee that the associated $X(\phi)$—which we may reasonably assume to be given by (4.4) also—will have finite second moments.

Since we must therefore abandon K_n in favour of some other version of N_n, let us list some of the properties required of such an N_n.

(N.1) $N_n \supset L_n$.

(N.2) The convergence to zero of the Lebesgue measures of a sequence of sets of fixed shape from $\mathscr{B}_b^n$ implies the convergence to zero of the sequence of indicator functions for these sets with respect to the topology of N_n.

(N. 3) For all $\phi \in N_n$, the functions $\tau_y\phi - \phi$ should converge to zero in N_n as $y \to 0$.

(N. 4) The topology of N_n is invariant under translations, and possibly under rotations as well.

(N. 5) Finally, we should like N_n to depend on some parameter, p say, such that for a particular value of p, which may as well be 1, N_n is identical with the space L_n^1. The motivation for this last requirement will become apparent in §6.

With agricultural applications in mind, (N. 1), (N. 2), and (N. 4) are fairly natural requirements. The remaining requirements are mainly for mathematical convenience, although (N. 3) is evidently related to (N. 2).

4.3 *N_n^p spaces.* For the above reasons, and also with a view to making use of theorem 4.1, we shall consider certain families N_n^p, $1 \leqq p \leqq 2$, of subspaces of L_n^1, which, in addition to possessing the properties (N. 1) to (N. 5) above are such that

(N. 6) each N_n^p is a Banach algebra (multiplication being convolution in L_n^1) obtained as the completion of K_n with respect to a certain norm, denoted by $\| \cdot \|_{[p]}$, which satisfies

$$\| \phi \|_{[p_1]} \leqq \| \phi \|_{[p_2]}$$

for all $\phi \in \mathscr{C}[(R_n, \mathscr{B}^n, \lambda^n)]$ whenever $1 \leqq p_1 < p_2 \leqq 2$.

5. The form of X, E_X, and C_X for $X \in G[N_n^p]$

The following are essentially preliminary consequences of the requirements set out in §4.3.

5.1 *Lemma.* $X \in G[N_n^p]$ if and only if there exist non-negative constants D_1 and D such that for all $\phi \in N_n^p$,

$$|E_X(\phi)| \leqq D_1 \| \phi \|_{[p]} \quad \text{and} \quad C_X(\phi, \phi) \leqq D \| \phi \|_{[p]}^2 .$$

We have then the following analogue of theorem 4.1:

5.2 *Lemma.* For every $X \in G[N_n^p]$ there exists (i) a measure F on $\mathscr{B}^n$ satisfying

$$\int |\tilde{\phi}(u)|^2 F(du) \leqq D \| \phi \|_{[p]}^2 \tag{5.1}$$

for some non-negative constant D and every $\phi \in N_n^p$; (ii) a complex constant c satisfying (4.2); and (iii) a completely additive random set function Z on $\mathscr{B}_b^n$ satisfying (4.3), such that the spectral representations (4.4), (4.5), and (4.6) hold.

Proof. The main step in the proof rests with establishing (4.6), since there is a well-trodden route (Itô, 1953; Yaglom, 1957, and p. 271 of Gel'fand & Vilenkin, 1964) leading from the spectral representation of C_X to that of X and E_X.

Suppose then that the functional C on $N_n^p \times N_n^p$ has the properties (C. 1) to (C. 5). The restriction of C to $K_n \times K_n$ then enjoys the same properties with respect to the topology of K_n, since convergence to zero in this topology implies convergence to zero in the topology of N_n^p. Consequently, from theorem 4.1, we have that there is a measure F with $k(F) < \infty$ such that the representation (4.6), and hence the condition (5.1), hold for all $\phi, \psi \in K_n$. Fixing $\psi \in K_n$ and subsequently fixing $\phi \in N_n^p$, the use of the Lebesgue monotone convergence theorem in conjunction with the fact that K_n is dense in N_n^p shows that this representation for C holds for all ϕ, $\psi \in N_n^p$.

5.3 *The N_n^p-spectral property.* We will say that F is an N_n^p-spectral measure, or has the N_n^p-spectral property, if it is a measure on $\mathscr{B}^n$ satisfying (5.1).

For a given N_n^p space it is of interest to find a reasonably simple condition which is sufficient, or better, both necessary and sufficient for a measure to have the N_n^p-spectral property. The basic result along these lines is the following:

5.4 *Theorem.* A measure F has the N_n^1-spectral property if and only if it is bounded.

Proof. Suppose F is bounded. From the properties of Fourier transforms of functions in $L_n^1 = N_n^1$ we have

$$\int |\tilde{\phi}(u)|^2 F(du) \leqq \|\tilde{\phi}\|_\infty^2 \int F(du) \leqq D \|\phi\|_1^2$$

where $D = F(R^n)$. Thus F is an N_n^1-spectral measure.

Conversely, if F is an N_n^1-spectral measure and the sequence $\delta_1, \delta_2, \cdots,$ is the approximate identity in L_n^1 given by

$$\delta_m(x) = (2\pi m^{-2})^{-n/2} \exp(-\tfrac{1}{2}m^2 |x|^2),$$

then $|\tilde{\delta}_m(u)| \uparrow 1$ for all $u \in R^n$ as $m \to \infty$ and

$$\int |\tilde{\delta}_m(u)|^2 F(du) \leqq D$$

for all m; so the boundedness of F follows from the monotone convergence theorem.

We conclude this section with two other obvious though useful results.

5.5 *Theorem.* The collection of N_n^p-spectral measures is closed under addition of measures, as well as under multiplication of measures by positive constants.

5.6 *Theorem.* We have $G[N_n^{p_1}] \subset G[N_n^{p_2}] \subset G[K_n]$ whenever $1 \leqq p_1 < p_2 \leqq 2$.

6. ORDINARY STATIONARY RANDOM PROCESSES ON R^n

6.1 *Definition.* By an ordinary stationary random process on R^n we mean a mapping $Y: R^n \to L^2(S,P)$ for some probability space $(S, \mathscr{E}, P)$ such that

(1) for all $x, y, u \in R^n$ we have that $E(Y(x)) = E(Y(y))$ and $(Y(x), Y(y)) = (Y(x+u), Y(y+u))$; and

(2) $\operatorname{Var}(Y(x) - Y(0)) \to 0$ as $x \to 0$ in R^n.

We recall that if Y is such an ordinary stationary random process then there exists (i) a bounded measure F on $\mathscr{B}^n$; (ii) a complex constant c satisfying (4.2), and (iii) a completely additive random set function Z on $\mathscr{B}_b^n$ satisfying (4.3) such that the spectral representations

$$Y(x) = \int \exp(iu \cdot x) Z(du), \tag{6.1}$$

$$E(Y(x)) = c, \text{ and} \tag{6.2}$$

$$(Y(x), Y(y)) \doteq \int \exp\{iu \cdot (x-y)\} F(du) \tag{6.3}$$

hold. Combining this result with 5.2 and 5.4 we see that by means of the identification

$$X(\phi) = \int \tilde{\phi}(u) Z(du)$$

for all $\phi \in L_n^1$, there is a natural one-to-one correspondence between ordinary stationary random processes on R^n and the members of $G[N_n^1]$.

7. Particular versions of N_n^p

The various examples of Segal algebras discussed in Reiter (1968, §1.5) suggest ways in which to construct specific N_n^p spaces. As a first version of N_n^p we consider the set of all $\phi \in \mathscr{C}[(R^n, \mathscr{B}^n, \lambda^n)]$ for which

$$\| \phi \|_{[p]} = \max[\| \phi \|_1, \| \phi \|_p] < \infty .$$

We shall denote this space by $L_n^{1,p}$ and write $\| \phi \|_{1,p}$ in place of $\| \phi \|_{[p]}$. In this case it is not too difficult to come up with conditions that F must necessarily satisfy to be an $L_n^{1,p}$-spectral measure. For example, F must satisfy the condition

$$\int |\tilde{\phi}(u)|^2 F(du) < \infty$$

for any $\phi \in L_n^{1,p}$ that we care to pick out. Pursuing this line with (4.1) in mind leads to the following

7.1 *Lemma.* If F is an $L_n^{1,p}$-spectral measure, then $k(F) \leqq n/p'$.

Proof. Consider the function

$$\kappa_k(x) = \pi^{-n/2} 2^{1-k} [\Gamma(\tfrac{1}{2}k)]^{-1} |2x|^{\frac{1}{2}(k-n)} K_{\frac{1}{2}(k-n)}(|x|),$$

where $K_t(s)$ is the Bessel function of order t with purely imaginary argument (Watson, 1944, p. 78). From §3.71 equation (2), §6.15 equation (4), §7.23 equation (1), and §15.7 of the same reference, it may be ascertained that $\kappa_k \in L_n^{1,p}$ whenever $n/p' < k < n$. Moreover, utilising theorem 40, p. 69 of Bochner & Chandrasekharan (1949), and the identity (2), §13.45 of Watson, we have

$$\tilde{\kappa}_k(u) = (1 + |u|^2)^{-k/2};$$

and this directly implies the stated conclusion.

On the other hand, the convexity of $\| \phi \|_r$ as a function of r (for fixed ϕ), the relation

$$\| \tilde{\phi} \|_{p'} \leqq (2\pi)^{n/p'} \| \phi \|_p,$$

and Hölder's inequality, together yield the following sufficient condition for a measure to have the $L_n^{1,p}$-spectral property.

7.2 *Lemma.* If F is either (i) bounded, or (ii) absolutely continuous with respect to Lebesgue measure, and $dF/d\lambda^n \in L_n^r$ for at least one $r \in [1, p/(2-p)]$, then F has the $L_n^{1,p}$-spectral property.

From this result and 5.5, it may be conjectured that for any $L_n^{1,p}$-spectral measure F there exists a decomposition of the form

$$F(A) = F_1(A) + \int_A \phi_2(u) du,$$

$A \in \mathscr{B}^n$, where F_1 is a bounded measure and $\phi_2 \geqq 0$ is in $L_n^{p/(2-p)}$. If we assume the truth of this conjecture then it appears that the widest class of processes possible for this family of N_n^p spaces, namely $G[L_n^{1,2}]$, will not be wide enough for our purposes; i.e., $G[L_n^{1,2}]$ will not net Fairfield Smith's law.

It is for this reason that a second version of N_n^p is considered, denoted by M_n^p, whose norm, $\|\cdot\|_{\{p\}}$, is defined for $\phi \in K_n$ as follows: We begin with the identity

$$\pi^{-n}[-\cos(\tfrac{1}{2}r\pi)]^n \int |\tilde{\phi}(u)|^2 \left| \prod_{j=1}^n u_j \right|^{r-2} du = [\Gamma(2-r)]^{-n} \int \phi * \phi^*(x) \left| \prod_{j=1}^n x_j \right|^{1-r} dx, \quad (7.1)$$

which holds for $1 < r < 2$, the common value of either side of which we shall denote b $[r, \phi]$. Further we define $[1, \phi]$ to be the value at $r = 1$ of the right-hand side of (7.1), the left-hand expression being similarly used to define $[2, \phi]$. Next we put

$$\|\phi\|_{(r)} = \sup_Q 2^{n(1-r)/r}[r, \phi_Q]^{\frac{1}{2}},$$

where Q denotes any $n \times n$ orthogonal matrix and ϕ_Q is defined by

$$\phi_Q(x) = \phi(Qx)$$

for all $x \in R^n$. Finally, for all $\phi \in K_n$, $\|\phi\|_{\{p\}}$ is given by

$$\|\phi\|_{\{p\}} = \max[\|\phi\|_1, \sup_{1 \leqq r \leqq p} \|\phi\|_{(r)}].$$

7.3 *Theorem.* The family of spaces M_n^p, $1 \leqq p \leqq 2$, satisfies the requirements (N. 1) to (N. 6) set out in sections 4.2 and 4.3.

Proof. It is a fairly straightforward matter to check that $\|\cdot\|_{\{p\}}$, as constructed, is a norm and to verify that the properties (N. 2), (N. 4), (N. 5), and (N. 6) are satisfied.

To check (N. 1) for $1 < p < 2$ we may proceed as follows: We start with the approximate identity $\delta_1, \delta_2, \cdots$, in L_n^1 given by

$$\delta_m(x) = \xi(x, m^{-1}) / \left\{ \int \xi(y, m^{-1}) dy \right\}$$

where

$$\xi(x, a) = \begin{cases} \exp[a^2/(|x|^2 - a^2)] & \text{for } |x| < a, \\ 0 & \text{otherwise.} \end{cases}$$

If for any $\phi \in L_n$ we put

$$\phi^{(m)} = \phi * \delta_m,$$

then $\phi^{(m)} \in K_n$ for every m, and $\|\phi - \phi^{(m)}\|_1 \to 0$ as $m \to \infty$. We wish to show that $\|\phi - \phi^{(m)}\|_{\{p\}} \to 0$. as $m \to \infty$.

To that end, put

$$B(\phi) = \max_{x \in R^n} |\phi(x)|$$

and let a_ϕ denote the radius of the smallest n-dimensional sphere outside of which ϕ vanishes. Then for every m

$$B(\phi^{(m)}) \leqq B(\phi) \text{ and } a_{\phi^{(m)}} \leqq a_\phi + m^{-1}.$$

Further, if for $a \geqq 0$ and $1 < r < 2$ we put

$$C(a, r) = \int_{|x| \leqq a} \left| \prod_{j=1}^{n} x_j \right|^{1-r} dx,$$

then for all $y \in R^n$ and every $n \times n$ orthogonal matrix Q we have

$$\int |\phi_Q(x-y)| \left| \prod_{j=1}^{n} x_j \right|^{1-r} dx \leqq B(\phi) C(a_\phi, r),$$

and hence

$$\left| \int \phi_Q * \psi_Q(x) \left| \prod_{j=1}^{n} x_j \right|^{1-r} dx \right| \leqq B(\phi) C(a_\phi, r) \| \psi \|_1 .$$

Noting that $C(a_1, r) < C(a_2, r)$ whenever $0 \leqq a_1 < a_2$ we may now show that

$$\| \phi - \phi^{(m)} \|_{(r)}^2 \leqq 2B(\phi)C(a_\phi + 1, r) \| \phi - \phi^{(m)} \|_1 .$$

Finally, the fact that

$$a^{1-r} \leqq 1 + a^{1-p}$$

for all $a \geqq 0$ and $1 \leqq r \leqq p$ allows the conclusion that

$$\| \phi - \phi^{(m)} \|_{\{p\}}^2 \leqq \| \phi - \phi^{(m)} \|_1^2 + 2B(\phi)C(a_\phi + 1, p) \| \phi - \phi^{(m)} \|_1 ;$$

and hence that $\| \phi - \phi^{(m)} \|_{\{p\}} \to 0$ as $m \to \infty$.

When $p = 2$ we may show similarly that

$$\| \phi - \phi^{(m)} \|_{\{2\}}^2 \leqq \| \phi - \phi^{(m)} \|_1^2 + 2B(\phi)C(a_\phi + 1, \tfrac{3}{2}) \| \phi - \phi^{(m)} \|_1 + 2^n \| \phi - \phi^{(m)} \|_2^2 ,$$

and the required result therefore follows from the fact that $\| \phi - \phi^{(m)} \|_2 \to 0$ as $m \to \infty$.

The above also provides the basis of an argument for establishing (N. 3).

We conclude the section by pointing out one fairly obvious comparison between $L_n^{1,p}$ and M_n^p spaces.

(7.4) *Theorem.* $G[M_n^2] \supset G[L_n^{1,2}]$

8. Fairfield Smith's law

In this section we are interested in R^2 only. To avoid subscripts we shall use (s, t) and (v, w) to denote points in R^2; and $A//d$ will denote the rectangle

$$|s| \leqq \tfrac{1}{2} e^d A^{\frac{1}{2}}, \ |t| \leqq \tfrac{1}{2} e^{-d} A^{\frac{1}{2}}$$

where $A > 0$ is the area of $A//d$, and $d \in R$ is a parameter representing the degree of departure from "squareness" of $A//d$. As in Section 1, γ will denote a parameter taking values in the interval $[1, 2)$ only.

8.1 *Theorem.* Let $X \in G[K_2]$ be a process whose spectral representation is determined by some complex constant c and the measure F on $\mathscr{B}^2$ given by

$$F(B) = |c|^2 \chi_B(0) + \{\sigma \Gamma(\gamma + 1) \sin(\tfrac{1}{2}\gamma\pi)/(2\pi)\}^2 \int_B |vw|^{1-\gamma} dv\, dw$$

for all $B \in \mathscr{B}^2$. Then X, or more precisely, the unique extension thereof, is in $G[M_2^p]$ for $p \geqq 3-\gamma$, and we have

$$\mathrm{Var}(X(\chi_{A//d})) = \sigma^2 A^{\gamma}; \tag{8.1}$$

in other words, $G[M_2^2]$ encompasses Fairfield Smith's law for all values of γ.

Proof. That F is an M_2^p-spectral measure for $p \geqq 3-\gamma$ follows from theorems 5.5 and 5.6 in conjunction with the definition of the M_2^p norm. The relation (8.1) may be established using the equations (4.5), (4.6), and (7.1).

The initial attraction of (8.1) as a means of accounting for various instances of agricultural variation is its simplicity. That it appears to give a satisfactory explanation of observations on many reported examples of such variation is also an attraction, but not in itself a compelling one. Reasons which explain just why such agreements with observations are likely to be encountered have been advanced by Whittle (1956).

However, it is clear that any process $X \in G[M_2^2]$ leading to the relation (8.1) will be, to some degree, less than satisfactory in those situations where an isotropic model is called for, and/or there is evidence that long thin plots are less variable than square plots of the same area. Our last theorem describes a possible alternative to Fairfield Smith's law which might be useful in such situations.

8.2 *Theorem.* Let $X \in G[K_2]$ be a process whose spectral representation is determined by some complex constant c and the measure F on $\mathscr{B}^2$ given by

$$F(B) = |c|^2 \chi_B(0) + \{\sigma^2 \Gamma(\gamma)/\Gamma(2-\gamma)\} \int_B (v^2 + w^2)^{1-\gamma} dv\, dw$$

for all $B \in \mathscr{B}^2$. Then X is in $G[L_2^{1,p}]$ whenever $p > 2/\gamma$ or $p = 2$, and is in $G[M_2^p]$ whenever $p \geqq 3-\gamma$. Further, if $A//S$ denotes a plot of area A and some given shape S, then $\mathrm{Var}(X(\chi_{A//S}))$ varies directly as A^{γ} when S is held fixed, while in the case of rectangles, $\mathrm{Var}(X(\chi_{A//d}))$, as a function of d for A fixed has a local maximum at $d = 0$.

Proof. That $X \in G[L_2^{1,p}]$ when $\gamma > 1$ and $2 \geqq p > 2/\gamma$ follows from an appropriate decomposition of F according to Lemma 7.2. The verifications of the remaining statements are more or less straightforward with the possible exception of the last one. For this we may use a Parseval-type relation to show that $\mathrm{Var}(X(\chi_{A//d}))$ is proportional to

$$A^{\gamma} \int_0^1 \int_0^1 (1-s)(1-t)(s^2 e^{2d} + t^2 e^{-2d})^{\gamma-2} ds\, dt,$$

and then carry out the indicated differentiations with respect to d under the integral signs.

We mention, in conclusion, that when this and other isotropic models were fitted to Fairfield Smith's data, they all exhibited in varying degrees the type of behaviour with respect to plot shape alluded to above, i.e., predicted that long thin plots would be less variable than square plots of the same area.

REFERENCES

BOCHNER, S. & CHANDRASEKHARAN, K. (1949). *Fourier Transforms.* Princeton University Press, Annals of Mathematics Studies, No. 19.

GEL'FAND, I. M. & SHILOV, G. E. (1964). *Generalized Functions, vol. I.* New York: Academic Press.

GEL'FAND, I. M. & VILENKIN, N. YA. (1964). *Generalized Functions, vol. IV.* New York: Academic Press.

Itô, K. (1953). Stationary random distributions. *Mem. Coll. Sci. Univ. Kyoto* **28,** 209–23.

Loève, M. (1963). *Probability Theory*. New York: Van Nostrand.

Matérn, B. (1960). *Spatial Variation*. Report of the Forest Research Institute of Sweden, vol. 49.

Reiter, H. (1969). *Classical Harmonic Analysis and Locally Compact Groups*. Oxford University Press.

Smith, H. Fairfield (1938). An empirical law describing heterogeneity in the yields of agricultural crops. *J. Agric. Sci.* **28,** 1–23.

Watson, G. N. (1944). *A Treatise on the Theory of Bessel Functions*. Cambridge University Press.

Whittle, P. (1956). On the variation of yield variance with plot size. *Biometrika* **43,** 337–43.

Whittle, P. (1962). Topographic correlation, power-law covariance functions, and diffusion. *Biometrika* **49,** 305–14.

Yaglom, A. M. (1957). Some classes of random fields in *n*-dimensional space related to stationary random processes. *Theory Prob. and Appl.* **2,** 273–320.

On the existence of a generalized shot-noise process

M. WESTCOTT
Imperial College, London

SUMMARY

This paper considers the existence of a generalization of the classical shot-noise model, in the sense of almost sure convergence of its formal defining series. This convergence may be conditional, absolute or as a Cauchy principal value. A very general necessary and sufficient condition is given and applied to several special cases, the most important being when the 'shots' are generated by a stationary Poisson process. The results are illustrated by examples. Extensions to a Poisson cluster process of 'shots' and sums of random measures are discussed.

1. INTRODUCTION

The basic model for a shot-noise process is very well known, with a history going back at least to Campbell in the first decade of the century. One important application has been to the voltage fluctuations in electron tubes, but there are numerous others; further references may be found in Smith (1958) and Daley (1971). In this paper, we shall consider a multivariate, multidimensional generalization of the model, along the lines of Daley (1971). Our principal concern is with the existence of the formally-defined random process, in the sense that it should be almost surely finite, though some distributional results for the process emerge as by-products, at least in the case when the 'shots' are generated by a Poisson process. The results obtained are a significant improvement on previous work.

In §2 of the paper we introduce the formal definition of our shot-noise process together with a few basic assumptions, and give the necessary preliminary distribution theory. Existence conditions for the process are derived in §3, both in general and in the most useful practical case when the 'shots' form a Poisson process. There are also several important corollaries. The conditions are illustrated in §4 by six examples. Section 5 treats the situation when the 'shots' are a Poisson cluster process, another case where explicit calculations are possible. The final section extends the shot-noise concept to sums of random measures.

2. DEFINITIONS AND PRELIMINARY RESULTS

Let $\{\boldsymbol{u}_j\}$ be a point process in $\mathscr{R}_n$, n-dimensional Euclidean space, with associated counting measure $N(\cdot)$. Suppose that $\{\boldsymbol{h}_j(\cdot)\}$ is a sequence of independent and identically distributed measurable random functions on $\mathscr{R}_n$, independent also of $\{\boldsymbol{u}_j\}$, which take values in $\mathscr{R}_m$ $(m, n \geqq 1)$. Then we may formally define a stochastic process $\boldsymbol{Y}(\boldsymbol{t})$ on $\mathscr{R}_m$ by

$$\boldsymbol{Y}(\boldsymbol{t}) = \sum_j \boldsymbol{h}_j(\boldsymbol{t} - \boldsymbol{u}_j). \tag{1}$$

This is, in essence, the formulation of Daley (1971). For readers unacquainted with the theory of point processes, an excellent contemporary survey may be found in Daley & Vere-Jones (1972).

Occasionally, it is convenient to write the sum as an integral, in the form

$$Y(t) = \int_{\mathscr{R}_n} h_u(t-u)N(du). \tag{2}$$

Here, $h_{u_j} \equiv h_j$ and we shall generally omit the range of integration.

Obviously, before we can use (1) or (2) as a model for some physical phenomenon and discuss its properties, we must ensure that the process is well-defined, in the sense that $Y(t)$ is almost surely (a.s.) finite for each t. The object of this paper is to derive and investigate conditions for the existence of the generalized shot-noise model in this sense.

Daley (1971) has previously given necessary and sufficient conditions for the existence of $Y(t)$ in the case when $N(\cdot)$ is a stationary Poisson process and h, nonrandom, has a particular symmetric form. He considered both the a.s. absolute convergence and convergence of the Cauchy principal value of (1). His conditions may be derived from ours as special cases, as we shall make very few initial assumptions about the nature of $N(\cdot)$ and h.

Specifically, let us work throughout with

Assumption A: If $B = \{u \in \mathscr{R}_n : \Pr(|h(u)| = \infty) > 0\}$, then $\Pr\{N(B) > 0\} = 0$.

Clearly, this assumption is essential to ensure that $Y(t)$ does not diverge trivially. One natural set of circumstances in which Assumption A will be satisfied is when $h(\cdot)$ is finite a.e. in $\mathscr{R}_n$ and $M(\cdot) = E\{N(\cdot)\}$ is absolutely continuous with respect to Lebesgue measure. Note that we use h as a generic random function with the common distribution of the $\{h_j\}$.

A complete description of the process $Y(t)$ is given by its *characteristic functional* (Bartlett, 1966, p.13), defined by

$$C_Y[\xi] = E\left[\exp\left\{i\int Y'(t)\xi(dt)\right\}\right],$$

where ξ is an $\mathscr{R}_m$-valued totally finite signed measure and $'$ indicates transpose. Now it is easy to see that

$$C_Y[\xi] = E\left(\prod_j E\left[\exp\left\{i\int h_j'(t-u_j)\xi(dt)\right\}\right]\right), \tag{3}$$

where the inner expectation refers to h, the outer expectation to the point process $\{u_j\}$ and we have used the independence of the two sequences. Equation (3) may be succinctly written as

$$C_Y[\xi] = G[\psi\{\xi(\cdot + u)\}], \tag{4}$$

where

$$\psi\{\xi(\cdot)\} = E\left[\exp\left\{i\int h(v)\xi(dv)\right\}\right]$$

and G is the probability generating functional of $N(\cdot)$ (for details of such functionals, see Westcott, 1972).

Temporarily setting aside the question of the existence of $Y(t)$, we remark that it is an immediate consequence of (4) that $Y(t)$ is stationary if $N(\cdot)$ is stationary. Also, despite the compact nature of (4) it is, unfortunately, well known that a closed-form expression for the generating functional G results only if $N(\cdot)$ is either the Poisson process or some close relative. In the Poisson case, we have

$$C_Y[\xi] = \exp\left(\int [\psi\{\xi(\cdot + t)\} - 1]\, M(dt)\right), \tag{5}$$

and the case of a Poisson cluster process will be treated later. The availability of an explicit form for C_Y makes the study of the existence of $Y(t)$ considerably simpler, but we shall treat this only as a special case, unlike Daley (1971).

3. Existence conditions

An obvious simplification of the whole problem is that, for $Y(t)$ to exist, it must exist component-wise. Thus without loss of generality we may take $m = 1$. Now, from (4), the characteristic function of $Y(t)$, for each $t \in \mathscr{R}_n$, is

$$\phi(\theta; t) = G[\psi(\theta; t - u)], \tag{6}$$

where

$$\psi(\theta; v) = E(e^{i\theta h(v)}).$$

For convenience, take $t = 0$ and change the sign of u to get

$$\phi(\theta) = G[\psi(\theta; u)].$$

This does not essentially change the problem and it is easy to refer our conditions back to the general format. Of course, if $Y(t)$ is stationary then it is sufficient to prove existence at just a single point t.

The first step in the argument is that $Y(t)$ exists iff

$$\lim_{\theta \to 0} \phi(\theta) = 1. \tag{7}$$

Then, since $|\psi(\theta; u)| \leqq 1$, dominated convergence implies that (7) is true if

$$\lim_{\theta \to 0} \int \log \psi(\theta; u)\, N(du) = 0 \quad \text{a.s..} \tag{8}$$

Now Assumption A says that $h(u)$ is bounded except for $u \in B$, which is a null set for $N(\cdot)$. Consequently, a standard argument lets us make ψ as close as we please to 1, uniformly in $u \in \mathscr{R}_n - B$, for small enough θ. Thus we may replace $\log \psi$ by $1 - \psi$ in (8) and so (7) holds iff

$$\lim_{\theta \to 0} \int \{1 - \psi(\theta; u)\}\, N(du) = 0 \quad \text{a.s..} \tag{9}$$

This is the most easily manipulated form of the existence condition. Let us write $\mathscr{I}(\theta)$ for the integral on the left side of (9).

We next define

$$F(x; u) = \Pr\{h(u) \leqq x\}, \qquad u \in \mathscr{R}_n,\ x \in \mathscr{R},$$

and set

$$H(x; u) = 1 - F(x; u) + F(-x; u), \qquad x \geqq 0,$$

$$K(x; u) = 1 - F(x; u) - F(-x; u), \qquad x \geqq 0.$$

Integration by parts in $1 - \psi$ readily shows that

$$1 - \psi(\theta; u) = \theta \int_0^\infty \sin(\theta x)\, H(x; u)\, dx - i\theta \int_0^\infty \cos(\theta x)\, K(x; u)\, dx, \tag{10}$$

and, after substitution of (10) in (9), we treat the real and imaginary parts of $\mathscr{I}(\theta)$ separately.

Real part: For an arbitrary $\delta > 0$, split the range of the $\boldsymbol{x}$-integration into $[0, \delta)$ and $[\delta, \infty)$, calling the two resulting double integrals I_1 and I_2, respectively. Thus

$$I_2 = \theta \int\int_{\delta}^{\infty} \sin(\theta x)\, H(x; \boldsymbol{u})\, dx\, N(d\boldsymbol{u}).$$

Now as H is monotone nonincreasing as x increases, application of the second mean value theorem (Titchmarsh, 1939, p. 379) to the inner integral shows that

$$\left|\theta \int_{\delta}^{\infty} \sin(\theta x)\, H(x; \boldsymbol{u})\, dx\right| \leqq 2H(\delta; \boldsymbol{u}).$$

Hence, assuming that

$$\int H(\delta; \boldsymbol{u})\, N(d\boldsymbol{u}) < \infty \text{ a.s. for some } \delta > 0, \tag{11}$$

and so for all greater δ, the dominated convergence theorem implies

$$I_2 \to 0 \text{ as } \theta \to 0. \tag{12}$$

Further,

$$I_1 = \theta \int\int_{0}^{\delta} \sin(\theta x)\, H(x; \boldsymbol{u}) dx\, N(d\boldsymbol{u}).$$

Suppose $|\theta| < \pi/2\delta$, so $|\theta x| < \frac{1}{2}\pi$; then Jordan's inequalities give

$$\frac{2\theta^2}{\pi} I_1^* \leqq I_1 \leqq \theta^2 I_1^*,$$

where

$$I_1^* = \int\int_0^{\delta} x H(x; \boldsymbol{u})\, dx\, N(d\boldsymbol{u}).$$

Thus $I_1 \to 0$ as $\theta \to 0$ iff

$$I_1^* < \infty \qquad \text{a.s.}. \tag{13}$$

But this implies $\int H(x; \boldsymbol{u})\, N(d\boldsymbol{u}) < \infty$ a.s. for all $x > 0$, since H is nonincreasing in x. Thus (13) $\Rightarrow$ (11) and, together with (12), we see that

$$\lim_{\theta \to 0} \operatorname{Re}\{\mathscr{I}(\theta)\} = 0 \Leftrightarrow I_1^* < \infty \text{ a.s. for arbitrary } \delta > 0. \tag{14}$$

Imaginary part: Again, split the range of x as before, for arbitrary $\delta > 0$, and call the integrals J_1 and J_2 respectively. Thus

$$J_2 = \theta \int\int_{\delta}^{\infty} \cos(\theta x)\, K(x; \boldsymbol{u})\, dx\, N(d\boldsymbol{u}).$$

Since K is the difference of two monotone functions we may apply the second mean value theorem to each part separately and, by an argument similar to the one above, prove that

$$J_2 \to 0 \text{ as } \theta \to 0 \text{ if (11) is true.} \tag{15}$$

Now if we define

$$J_1^* = \int\int_0^\delta K(x; \boldsymbol{u})dx\, N(d\boldsymbol{u}),$$

then

$$|J_1 - \theta J_1^*| \leqq \theta \int\int_0^\delta (1 - \cos\theta x)\, H(x; \boldsymbol{u})dx\, N(d\boldsymbol{u})$$

$$\leqq \tfrac{1}{2}\theta^3\delta \int\int_0^\delta xH(x; \boldsymbol{u})\, dx\, N(d\boldsymbol{u})$$

$$= \tfrac{1}{2}\theta^3\delta I_1^*. \tag{16}$$

Thus from the implication after (13), together with (14), (15) and (16), we deduce that

$$I_1^* < \infty \text{ a.s.}, |J_1^*| < \infty \text{ a.s.} \Rightarrow \lim_{\theta\to 0} \mathscr{I}(\theta) = 0.$$

Conversely, if (9) is true then (14) implies that $I_1^* < \infty$ a.s. and so (16) implies that $J_1^* < \infty$ a.s.. Thus:

The conditions $I_1^* < \infty$ *a.s.*, $|J_1^*| < \infty$ *a.s., for arbitrary* $\delta > 0$, *are necessary and sufficient for* (9), *hence sufficient for* (7).

The necessity of these conditions follows easily in the same way as in Westcott (1971). Consider the set of $N(\cdot)$ for which these integrals diverge, $\mathscr{D}$ say. If $\Pr(\mathscr{D}) = p > 0$, it is clear from the above that the integrals in (9) and (8) diverge exactly when $N \in \mathscr{D}$, and hence $|\phi(\theta)| \leqq 1 - p$. And this contradicts (7).

These conclusions may be stated as

Theorem 1.

Under Assumption A, the necessary and sufficient conditions for the existence at $\boldsymbol{t} = \boldsymbol{0}$ of the process $\boldsymbol{Y}(\boldsymbol{t})$ defined formally by (1) are

$$\int\int_0^\delta xH(x; \boldsymbol{u})dx\, N(d\boldsymbol{u}) < \infty \text{ a.s. for arbitrary } \delta > 0, \tag{17}$$

$$\left|\int\int_0^\delta K(x; \boldsymbol{u})\, dx\, N(d\boldsymbol{u})\right| < \infty \text{ a.s. for arbitrary } \delta > 0. \tag{18}$$

For general $\boldsymbol{t}$, replace $\boldsymbol{u}$ by $\boldsymbol{u} + \boldsymbol{t}$ in both H and K.

Remark 1.1. This result is completely general; the only extraneous assumption merely excludes trivial cases. However, this extreme generality means that conditions (17), (18) are perhaps impracticable, especially as they are stochastic in nature (cf. the existence conditions in Westcott, 1971). Consequently, it is of interest to examine special cases. The most useful one, from all points of view, is when $N(\cdot)$ is a stationary Poisson process; this will be the content of Theorem 2. Other cases occur in the following Corollaries—note that some which could be associated with Theorem 1 are in fact taken with Theorem 2 for ease of exposition.

Remark 1.2. For an enlightening alternative expression of the existence conditions, via truncation, see Corollary 2.2 and its preamble.

Remark 1.3. The conditions (17), (18) ensure the a.s. *conditional* convergence of the formal series (1), a case not treated in Daley (1971). If we allow the integrals to exist merely as Cauchy principal values, that is $\lim_{R\to\infty}\int_{|\boldsymbol{u}|\leq R}$ finite, then $\boldsymbol{Y}(\boldsymbol{t})$ exists as a Cauchy principal value for some $\boldsymbol{t}$. The case of *absolute* convergence is treated in Corollary 1.1. Example 1 illustrates these three modes of convergence.

Corollary 1.1. The series (1) is a.s. absolutely convergent iff

$$\int\int_0^\delta H(x;\boldsymbol{u})\,dx\,N(d\boldsymbol{u})<\infty \text{ a.s. for arbitrary } \delta>0. \tag{19}$$

This is clear from the facts that $H = K$ for nonnegative h and that the integral in (18) then dominates the one in (17). Let us say in future that $\boldsymbol{Y}(\boldsymbol{t})$ exists absolutely in this case.

For a useful special case which is intermediate between Theorems 1 and 2, suppose $N(\cdot)$ is stationary and ergodic. This implies that

$$|A|^{-1}N(A)\to\lambda \text{ a.s. as } |A|\to\infty,$$

where $A\subset\mathscr{R}_n$, $|A|$ is Lebesgue measure of A and λ is a finite constant. Label the points $\{\boldsymbol{u}_j\}$ of $N(\cdot)$ in order of increasing distance from the origin to get a new point process $\{v_j\}$ on $[0,\infty)$, for which

$$t^{-n}N(0,t)\to\lambda_1 \text{ a.s. as } t\to\infty;\ \lambda_1=\frac{\pi^{\frac{1}{2}n}}{\Gamma(1+\frac{1}{2}n)}\lambda.$$

Therefore, given $\varepsilon>0$, we have $(\lambda_1-\varepsilon)<t^{-n}N(0,t)<(\lambda_1+\varepsilon)$ for t sufficiently large, i.e. taking $t=v_r$,

$$\left(\frac{r}{\lambda_1+\varepsilon}\right)^{1/n}<v_r<\left(\frac{r}{\lambda_1-\varepsilon}\right)^{1/n} \text{ for } r \text{ sufficiently large.}$$

Thus if both H and K are ultimately monotone in $|\boldsymbol{u}|$, since we may ignore a bounded subset of $\mathscr{R}_n$ in considering convergence of (17), (18), it is clear that (17) is true iff

$$\sum_{j\geq 0}\int_0^\delta x\,H(x;\lambda_1^{-1}j^{1/n})\,dx<\infty,$$

with a similar expression corresponding to (18). Application of Cauchy's integral test, as in Daley (1971), leads to

Corollary 1.2. When $N(\cdot)$ is stationary and ergodic with finite mean, and both H and K are ultimately monotone in $|\boldsymbol{u}|$, the process $\boldsymbol{Y}(\boldsymbol{t})$ exists iff

$$\int_0^\infty\int_0^\delta x\,H(x;y)\,y^{n-1}\,dxdy<\infty \text{ for arbitrary } \delta>0, \tag{20}$$

$$\left|\int_0^\infty\int_0^\delta K(x;y)\,y^{n-1}\,dxdy\right|<\infty \text{ for arbitrary } \delta>0. \tag{21}$$

An example in which the conditions on H and K are satisfied is the problem treated in Daley (1971). It seems difficult to extend the corollary to other less restrictive forms of monotonicity for H and K, since we need the order properties associated with $\mathscr{R}$ but not $\mathscr{R}_n$, $n > 1$, in the derivation. Note that an assumption that H and K are ultimately of bounded variation leads to the same result, since such functions can be written as the difference of two monotone functions.

By far the most common situation in discussions of shot-noise models is when $N(\cdot)$ is a stationary Poisson process, and in this case the results simplify significantly; this parallels similar simplifications in the study of cluster point processes (Westcott, 1971). For, from (5) and (6), the characteristic function of $Y(\boldsymbol{t})$ is

$$\phi(\theta; \boldsymbol{t}) = \exp\left[\lambda \int \{\psi(\theta; \boldsymbol{u}) - 1\} d\boldsymbol{u}\right],$$

and it is obvious that (7) is true, *for all* $\boldsymbol{t}$, iff

$$\lim_{\theta \to 0} \int \{\psi(\theta; \boldsymbol{u}) - 1\} d\boldsymbol{u} = 0.$$

By exact analogy with the arguments used to establish (9) and (19), we have

Theorem 2.

When $N(\cdot)$ is a stationary Poisson process and $\boldsymbol{h}(\cdot)$ is finite a.e. (so that Assumption A holds), the process $\boldsymbol{Y(t)}$ defined formally by (1) exists iff

$$\int\int_0^\delta x\, H(x; \boldsymbol{u})\, dx d\boldsymbol{u} < \infty \quad \text{for arbitrary } \delta > 0, \tag{22}$$

$$\left|\int\int_0^\delta K(x; \boldsymbol{u})\, dx d\boldsymbol{u}\right| < \infty \quad \text{for arbitrary } \delta > 0. \tag{23}$$

It exists as a Cauchy principal value iff these integrals exist as Cauchy principal value integrals, with respect to $\boldsymbol{u}$. The process exists absolutely iff

$$\int\int_0^\delta H(x; \boldsymbol{u})\, dx d\boldsymbol{u} < \infty \qquad \text{for arbitrary } \delta > 0. \tag{24}$$

These conditions are a substantial improvement on those of Daley (1971), in that h is left practically arbitrary. If in fact h takes his special form we get the conditions (20), (21) without, however, any need for assumptions of ultimate monotonicity. Note, incidentally, that Daley's necessary condition for convergence (equation (2) of Daley, 1971) is not quite correct, since it implies, for example, that $|h(\boldsymbol{u})| \to 0$ as $|\boldsymbol{u}| \to \infty$ is a necessary condition for integrability of h.

An interesting consequence of the theorem is

Corollary 2.1. If $h(\boldsymbol{u})$ is a symmetric random variable for each $\boldsymbol{u}$, the process $\boldsymbol{Y(t)}$ exists iff

$$\int\int_0^\delta x\, H(x; \boldsymbol{u})\, dx d\boldsymbol{u} < \infty \text{ for arbitrary } \delta > 0.$$

Thus an assumption of symmetry means we need less restrictions on H to ensure convergence; see Example 3.

There is another way of writing the existence conditions which is, perhaps, easier to appreciate.

Let

$$h_\delta(\boldsymbol{u}) = \begin{cases} h(\boldsymbol{u}) & \text{if } |h(\boldsymbol{u})| \leqq \delta \\ \delta & \text{if } h(\boldsymbol{u}) > \delta \\ -\delta & \text{if } h(\boldsymbol{u}) < -\delta. \end{cases}$$

That is, we truncate the function h in a manner reminiscent of other areas of probability theory, though the usual truncation is to zero rather than δ. It is now clear, on integration by parts, that

$$2\int_0^\delta xH(x;\boldsymbol{u})dx = E[\{h_\delta(\boldsymbol{u})\}^2],$$

$$\int_0^\delta K(x;\boldsymbol{u})dx = E\{h_\delta(\boldsymbol{u})\}.$$

Hence the existence conditions (22), (23) become

$$\int E[\{h_\delta(\boldsymbol{u})\}^2]d\boldsymbol{u} < \infty \quad \text{for arbitrary } \delta > 0, \tag{25}$$

$$\left|\int E\{h_\delta(\boldsymbol{u})\}d\boldsymbol{u}\right| < \infty \quad \text{for arbitrary } \delta > 0, \tag{26}$$

while (24) gives, for nonnegative h, the condition

$$\int E\{h_\delta(\boldsymbol{u})\}d\boldsymbol{u} < \infty \quad \text{for arbitrary } \delta > 0. \tag{27}$$

Further, in the important special case of bounded h, we have

Corollary 2.2. If $h \geqq 0$ and a.s. $\sup h(\boldsymbol{u}) < \infty$, then $\boldsymbol{Y}(t)$ exists iff

$$\int E\{h(\boldsymbol{u})\}d\boldsymbol{u} < \infty. \tag{28}$$

If h is nonrandom the condition reduces to integrability of h.

Remark 2.1. Clearly, similar results apply in the case of Theorem 1 and its corollaries, and need no statement.

Remark 2.2. The condition (27) is given by Itô (1972).

Remark 2.3. Equation (28) is exactly the sufficient condition for existence we obtain in general by use of the relation $E\{\boldsymbol{Y}(\boldsymbol{t})\}$ finite $\Rightarrow$ $\boldsymbol{Y}(\boldsymbol{t})$ a.s. finite. It is interesting that the condition is also necessary if h is bounded. The nonnegativity is essential here; if $n = 1$ and $h(u) = \sin u / u^{\frac{1}{2}}$ for $u > 0$, then (28) is satisfied yet (25) is not.

Remark 2.4. Perhaps not surprisingly, there is a striking analogy between (25), (26) and the three-series theorem of Kolmogorov (Loève, 1963, p. 237) for the convergence of sums of independent random variables. Now the terms in (1) are never actually independent, so a direct application of the three-series theorem is not possible. However, we can apply it *formally* to derive the existence conditions when $N(\cdot)$ is Poisson. For we may envisage (1) as a 'sum' of uncountably many terms,

$$\sum_u h_u(t-u)\,N(du),$$

and when $N(\cdot)$ is Poisson, with independent increments, the summands are independent. A direct substitution into the conditions of the three-series theorem leads, after a little manipulation, to (25) and (26); there is no extra term because of our different mode of truncation.

4. EXAMPLES

In these illustrative examples of the above results, we shall only consider the case when $N(\cdot)$ is a Poisson process, i.e. Theorem 2 and its corollaries apply, and $n = 1$. Generally, we shall also take h to be nonnegative. In this case, it is worth noting that (24) or (26) is equivalent to

$$A(x) = \int_{-\infty}^{\infty} \Pr\{h(u) > x\}\,du \in L(0,\delta) \text{ for arbitrary } \delta > 0. \tag{29}$$

Example 1: To show the differences between a.s. covergence of (1) as a Cauchy principal value, conditionally and absolutely, consider the nonrandom function

$$\left.\begin{aligned} h(u) &= \frac{\sin u^{\frac{1}{3}}}{1+u^a} & u \geqq 0 \\ &= -h(-u) & u < 0 \end{aligned}\right\}.$$

The three modes of convergence can be obtained by choosing $a = 2/3$, $a = 1$ and $a = 4/3$, respectively.

Example 2: Let $h(u) = Xg(u)$, $g \geqq 0$, $X \geqq 0$ a random variable. Then

$$A(x) = \int_0^{\infty} \Pr\{X > x/g(u)\}du.$$

Now without loss of generality we may choose g to be monotone decreasing and nonzero for $u \geqq 0$, since we can rearrange g otherwise without changing the value of the integral (see Hardy, Littlewood & Polya, 1934, §10.12). So

$$\int_0^{\delta} A(x)dx = \int_0^{\infty} g(u)\left\{\int_0^{\delta/g(u)} \Pr(X > y)dy\right\}du. \tag{30}$$

Suppose now that

$$\Pr(X > y) = y^{-\alpha}L(y),$$

where L is a slowly varying function and $\alpha \geqq 0$.

$\alpha > 1$: The inner integral in (30) is bounded, so

$$(29) \Leftrightarrow \int_0^{\infty} g(u)du < \infty.$$

More generally, this is true whenever $E(X) < \infty$.

$0 < \alpha < 1$: By Feller (1971, p. 281, Theorem 1) and (30),

$$(29) \Leftrightarrow \int_0^{\infty} \{g(u)\}^{\alpha} L[\{g(u)\}^{-1}]\,du < \infty$$

$\alpha = 1$: We cannot make any general statement about the validity of (29) without specifying L more precisely. For example,

$$L \sim \text{constant}: (29) \Leftrightarrow \int_0^\infty g(u)\,|\log g(u)|\,du < \infty,$$

$$L \sim (\log u)^{-1}: (29) \Leftrightarrow \int_0^\infty g(u)\,\big|\log|\log g(u)|\big|\,du < \infty.$$

Example 3: Let $h(u) = Xg(u)$, $g \geqq 0$, X a symmetric random variable. Then by Corollary 2.1, existence is equivalent to

$$\int_0^\delta x \int_{-\infty}^\infty \Pr\{|h(u)| > x\}\, du dx < \infty \text{ for arbitrary } \delta > 0. \tag{31}$$

As in Example 2, this becomes

$$2\int_0^\infty \{g(u)\}^2 \left\{\int_0^{\delta/g(u)} y\Pr(X > y)dy\right\} du,$$

where, without loss of generality, $g(u)$ is decreasing.

Again, assume $\Pr(X > y)$ is regularly varying of index $\alpha \geqq 0$ and apply Feller's theorem.

$0 \leqq \alpha < 2$: When $L \sim$ constant, we have

$$(31) \Leftrightarrow \int_0^\infty \{g(u)\}^\alpha du < \infty,$$

with an obvious modification in the more general case.

$\alpha > 2$: The inner integral of (31) is bounded, so

$$(31) \Leftrightarrow \int_0^\infty \{g(u)\}^2 du < \infty.$$

This is generally true whenever $E(X^2) < \infty$.

Again, when $\alpha = 2$ we must specify L more precisely to get results.

This should be contrasted with Example 2, which is its absolute convergence equivalent. It is, perhaps, surprising that when $0 \leqq \alpha < 1$ the conditions for absolute and conditional convergence of (1) are identical; however, we require less of g for $\alpha \geqq 1$.

Note that all the conclusions of both examples are still valid if X is replaced by $X(u)$, a stationary random process. See also the next example.

Example 4: Let $h(u) = X(u)g(u)$, where $X(u)$ is a symmetric stable process on $(0, \infty)$ of index $\alpha > 1$ and $g \geqq 0$ is decreasing and nonzero only for $u \geqq 0$ (again, without loss of generality). Using the result that $X(u)$ is distributed as $Xu^{1/\alpha}$, where X is a symmetric stable random variable of index α, we find from Examples 2 and 3 that

$$Y(t) \text{ exists} \Leftrightarrow \int_0^\infty \{g(u)\}^2\, u^{2/\alpha} du < \infty$$

$$Y(t) \text{ exists absolutely} \Leftrightarrow \int_0^\infty g(u) u^{1/\alpha} du < \infty.$$

The case $\alpha = 1$ would again require special treatment. When $0 < \alpha < 1$, so that X is nonnegative, we come back to Example 2 with function $u^{1/\alpha}g(u)$ and $L \sim$ constant (for this last result, see Feller, 1971, p. 448).

Example 5: Let $h(u) = g\{|X(u)|\}$, where $X(u)$ is a symmetric stable process of index α on $(0, \infty)$ and $g \geqq 0$ is decreasing. Hence the inverse of g exists; call it γ. Then it is easy to show that

$$A(x) = \{\gamma(x)\}^{\alpha} \int_0^{\infty} y^{-(1+\alpha)} \Pr\{|X| < y\}\, dy.$$

However, $\Pr\{|X| < y\} \sim Cy$, C constant, as $y \to 0$ if $\alpha \geqq 1$, while it is smaller than any power of y if $\alpha < 1$; see Ibragimov & Linnik (1971, pp. 61–2). Thus the process $\boldsymbol{Y}(t)$ will not exist if $\alpha \geqq 1$ and exists for $\alpha < 1$ if and only if $\int_0^{\delta}\{\gamma(x)\}^{\alpha} dx < \infty$, i.e.

$$\int_c^{\infty} u^{\alpha - 1} g(u)\, du < \infty$$

for some $c > 0$.

The process does not exist for $\alpha \geqq 1$ because $X(u)$ oscillates too fast.

Example 6: Let $h(u) = g\{X(u)\}$, where $X(u)$ is a nonnegative integer-valued random process such that $X(0) = 0$ and $X(u) \to \infty$ a.s. as $u \to \infty$, and g is bounded, positive and decreasing to zero (if we let g be unbounded, h will be infinite on sets of nonzero measure). Define γ as in the previous example.

Now as $X(u)$ is constant over intervals, we see that

$$\begin{aligned}\int_0^{\delta} A(x)dx &= \sum_{n=c}^{\infty} \{g(n) - g(n+1)\} \int_0^{\infty} \Pr\{X(u) < n\}\, du, \\ &\qquad\qquad c = [\gamma(\delta)] + 1, \\ &= \sum_{n=c}^{\infty} g(n+1) \int_0^{\infty} \Pr\{X(u) = n\}\, du + c_1.\end{aligned}$$

Thus $\boldsymbol{Y}(t)$ exists iff

$$\sum_{n=c}^{\infty} g(n+1) \int_0^{\infty} \Pr\{X(u) = n\}\, du < \infty \text{ for some } c > 0, \tag{32}$$

where the integral is the expected sojourn time of $X(u)$ in state n.

In particular, take $X(u)$ to be a birth-death process (Cox & Miller, 1965, p. 175), with parameters

$$\lambda_0 > 0,\quad \mu_0 = 0;\quad \lambda_i > 0,\quad \mu_i \geqq 0 \qquad (i = 1, 2, \cdots).$$

Let

$$w_0 = 1,\quad w_n = \frac{\lambda_0 \cdots \lambda_{n-1}}{\mu_1 \cdots \mu_n} \qquad (n = 1, 2, \cdots),$$

and assume

$$\Sigma(\lambda_n w_n)^{-1} < \infty \tag{33}$$

so that $X(u)$ is transient. Now, using either the forward equations of the process or the elegant spectral theory of Karlin and McGregor (Cox & Miller, 1965, p. 186), it can be shown that

$$\int_0^\infty \Pr\{X(u) = n\}\, du = w_n \sum_{j=n}^{\infty} (\lambda_j w_j)^{-1}.$$

Hence (32) can be investigated under various assumptions about the parameters. For example, suppose

$$\lambda_0 = a,\ \lambda_n = \lambda n^\alpha + a,\ \mu_n = \mu n^\alpha + b, \qquad (n = 1, 2, \cdots).$$

(i) $\lambda > \mu$: Here

$$w_n \sim c_2 \left(\frac{\lambda}{\mu}\right)^n n^{\alpha(a/\lambda - b/\mu - 1)},$$

so that

$$w_n \sum_{j=n}^{\infty} (\lambda_j w_j)^{-1} \sim c_3 n^{-\alpha},$$

i.e.

$$(32) \Leftrightarrow \sum_{n=c}^{\infty} g(n) n^{-\alpha} < \infty \Leftrightarrow \int_c^\infty g(u) u^{-\alpha} du < \infty.$$

So if $\alpha > 1$, $Y(t)$ always exists.

(ii) $\lambda = \mu$, $\alpha \leqq 1$: (if $\alpha > 1$, (33) does not hold)

(a) $\alpha < 1$, $a > b$: It is easy to show that

$$w_n \lambda_n \sim c_4 \exp\{(a - b) n^{1-\alpha}\},$$

whence

$$(32) \Leftrightarrow \sum_{n=c}^{\infty} g(n) n^{-\alpha} < \infty, \text{ as before}.$$

(b) $\alpha = 1$, $a > b + 1$: Again, elementary calculations show

$$w_n \lambda_n \sim c_5 n^{(a-b)},$$

so that

$$(32) \Leftrightarrow \sum_{n=c}^{\infty} g(n) < \infty \Leftrightarrow \int_c^\infty g(u) du < \infty.$$

5. Shot noise generated by a Poisson cluster process

It has previously been mentioned that specific distributional results for $Y(t)$ are not available unless the probability generating functional, G, of $N(\cdot)$ is expressible in closed form. In such cases we must be content with the generalities of Theorem 1, for example. When $N(\cdot)$ is Poisson we have seen that more satisfying conclusions are possible (Theorem 2). We now look briefly at the case of $N(\cdot)$ a stationary Poisson cluster process. This is a plausible model in some circumstances, and it also provides an example when $Y(t)$ exists although $M(\cdot)$ is infinite on bounded sets, a situation not hitherto encountered.

For details of the basic Poisson clustering model we refer to Westcott (1971) and Daley & Vere-Jones (1972). We have

$$G[\xi] = \exp\left[-\lambda \int \{1 - G_2[\xi \mid t]\} dt\right],$$

where G_2 is the probability generating functional of a cluster centred at $\boldsymbol{t}$. Thus conditions for existence will involve h and the cluster structure.

For simplicity, let us stick to the case $n = 1$, $h \geqq 0$, and work with Laplace transforms rather than characteristic functions. From (6),

$$E(e^{-\theta Y(t)}) = \exp\left[-\lambda \int_{-\infty}^{\infty} \{1 - G_2[\psi(\theta;\cdot)\,|\,u]\}du\right],$$

where

$$\psi(\theta; v) = E(e^{-\theta h(v)}), \quad \theta \geqq 0, \quad v \in \mathscr{R},$$

so that $\boldsymbol{Y}(t)$ exists iff

$$\mathscr{I}^*(\theta) = \int_{-\infty}^{\infty} \{1 - G_2[\psi(\theta;\cdot)\,|\,u]\}du < \infty \text{ for some } \theta > 0. \tag{34}$$

Now suppose there are ν points in the cluster, at $\tau_1, \cdots, \tau_\nu$; ν is a.s. finite of course. Then

$$\begin{aligned} 0 \leqq 1 - G_2[\psi(\theta;\cdot)\,|\,u] &= E\left\{1 - \prod_{j=1}^{\nu} \psi(\theta; u + \tau_j)\right\} \\ &\leqq E\left[\sum_{j=1}^{\nu} \{1 - \psi(\theta; u + \tau_j)\}\right] \\ &= \int_{-\infty}^{\infty} \{1 - \psi(\theta; u + \tau)\} M_s(d\tau), \end{aligned}$$

where $M_s(\cdot)$ is the first moment measure for the cluster. Thus

$$\mathscr{I}^*(\theta) \leqq E(\nu) \int_{-\infty}^{\infty} \{1 - \psi(\theta; u)\}\, du,$$

since $E(\nu) = M_s(\mathscr{R})$, so (34) holds if

$$E(\nu) < \infty, \int_{-\infty}^{\infty} \{1 - \psi(\theta; u)\} du < \infty \text{ for some } \theta > 0.$$

In addition

$$E\left\{1 - \prod_{j=1}^{\nu} \psi(\theta; u + \tau_j)\right\} \geqq E\{1 - \psi(\theta; u + \tau_1)\},$$

so that

$$\mathscr{I}^*(\theta) \geqq \int_{-\infty}^{\infty} \{1 - \psi(\theta; u)\}\, du.$$

That is, if $E(\nu) < \infty$, the convergence of $\int_{-\infty}^{\infty} \{1 - \psi(\theta; u)\}du$ is necessary and sufficient for existence of $\boldsymbol{Y}(t)$. Since this integral is basically the same as (9), conditions for its finiteness have already been examined.

Note that under the additional hypothesis that $\psi(\theta; u)$ is an increasing function as $u \to \pm\infty$, this conclusion is also deducible from Corollary 1.2, since $E(\nu) < \infty$ implies $M(\cdot)$ finite, and Poisson cluster processes are ergodic (Westcott, 1971).

Suppose, now, that $\psi(\theta; u) = 1$ for $u < 0$ and increases monotonically as $u \to \infty$. Then

$$1 - \prod_{j=1}^{\nu} \psi(\theta; u + \tau_j) \leqq 1 - \{\psi(\theta; u + \tau_1)\}^\nu,$$

so that

$$\mathscr{I}^*(\theta) \leqq \int_0^\infty \{1 - P[\psi(\theta; u)]\}du,$$

where $P(z) = E(z^\nu)$, $|z| \leqq 1$. Thus the convergence of this integral is sufficient for (34). A particular case comes from

$$1 - z^\nu \leqq \nu^\alpha(1-z)^\alpha, \qquad 0 < \alpha \leqq 1,$$

whence

$$\int_0^\infty \{1 - P[\psi(\theta; u)]\}du \leqq E(\nu^\alpha) \int_0^\infty [1 - \psi(\theta; u)]^\alpha du. \tag{35}$$

Thus the finiteness of both quantities on the right-hand side of (35) is sufficient for existence of $\mathbf{Y}(t)$. The interesting point here is that if $\alpha < 1$ we could have $E(\nu) = \infty$, whence $M(I) = \infty$ for bounded I, yet the process $\mathbf{Y}(t)$ still exists.

We summarize the various conclusions in

Theorem 3.

Suppose $\mathbf{Y}(t)$ is defined by (1) when $n = 1$, $N(\cdot)$ is a stationary Poisson cluster process and h is nonnegative and finite a.e.. Let $\psi(\theta; v) = E(e^{-\theta h(v)})$, ν be the random cluster size and $P(z) = E(z^\nu)$. Then,

(i) if $E(\nu) < \infty$, $\mathbf{Y}(t)$ exists iff

$$\int_0^\infty \{1 - \psi(\theta; u)\}\, du < \infty;$$

(ii) if $\psi(\theta; u) \equiv 1$ for $u < 0$ and is monotone nondecreasing as $u \to \infty$, $\mathbf{Y}(t)$ exists if

$$\int_0^\infty \{1 - P[\psi(\theta; u)]\}\, du < \infty.$$

In the case $E(\nu^\alpha) < \infty$, $0 < \alpha < 1$, the process exists if $\int_0^\infty \{1 - \psi(\theta; u)\}^\alpha du < \infty$.

6. SUMS OF RANDOM MEASURES

In a problem posed to the 1971 Conference on Point Processes, R. Pyke asked for the extension of Daley's results to sums of the form

$$\nu(S) = \sum_j \mu_j(S - \mathbf{t}_j), \tag{36}$$

where $\{\mu_j\}$ are possibly random measures and S a set in $\mathscr{R}_m$, say. In this final section we shall investigate such extensions. As usual, the $\{\mu_j\}$ will be taken independent and identically distributed, with generic element μ.

Firstly, if μ is just an arbitrary set function, the previous work is completely applicable for each S. However, as soon as μ is made finitely or completely additive, this additional structure naturally imposes some structure on the sum, i.e. on $\nu(S)$. One important implication of this structure is that, since a sum of μ-values over disjoint sets must be *absolutely* convergent, little will be lost if we take μ to be nonnegative, i.e. a measure, as in Pyke's formulation of the problem. Then ν is clearly also a measure.

Further, we shall assume μ is a.s. a Borel measure, defined on the Borel sets of $\mathscr{R}_m$, to which S will always belong. Thus μ is finite on compact (closed bounded) sets S, and for such S the summands in (36) must always be finite. This stands in contrast with possibilities for (1). Of course, $\mu(S+t)$ may be unbounded as $t \to \pm\infty$; we shall see later that convergence of (36) is still possible in this case. In summary, then, we are restricting the generality of the problem slightly, though not in any essential way. The question of (absolute) convergence of (36) for all bounded Borel sets S may now be studied exactly as before. Assume that $\{\boldsymbol{t}_j\}$ is a stationary Poisson process in $\mathscr{R}_m$. Then it is easily shown that the necessary and sufficient condition for a.s. convergence is

$$\int [1 - E\{e^{-\theta\mu(S+\boldsymbol{t})}\}]d\boldsymbol{t} < \infty \text{ for some } \theta > 0,$$

which reduces readily to the equivalent forms

$$\int \Pr\{\mu(S+\boldsymbol{t}) > x\}\, d\boldsymbol{t} \in L(0,\delta) \text{ for arbitrary } \delta > 0$$

or

$$\int E\{\mu_\delta(S+t)\}dt < \infty \text{ for arbitrary } \delta > 0. \tag{37}$$

Here, $\mu_\delta(\cdot) = \min\{\mu(\cdot),\delta\}$. Extension of these results to all the other cases considered in §3 is immediate.

Equation (37) has an attractive alternative when $\mu(S+\boldsymbol{t})$ is bounded in $\boldsymbol{t}$ for fixed S. For then, by Fubini's theorem,

$$\int E\{\mu_\delta(S+\boldsymbol{t})\}\, d\boldsymbol{t} = |S|\; E\{\mu(\mathscr{R}_m)\},$$

so that (37) holds iff $E\{\mu(\mathscr{R}_m)\} < \infty$. In particular, if μ is nonrandom it must be a totally finite measure.

These conclusions are contained in

Theorem 4.

Suppose that in the formal sum (36) the μ_j are independent and identically distributed Borel measures, $\{\boldsymbol{t}_j\}$ is a stationary Poisson process in $\mathscr{R}_m$ and S a bounded Borel set in $\mathscr{R}_m$. Then the series converges a.s. iff (37) holds. If, for fixed S, a.s. $\sup_{\boldsymbol{t}} \mu(S+\boldsymbol{t}) < \infty$, then (37) holds iff $E\{\mu(\mathscr{R}_m)\} < \infty$.

The Laplace transform of $\nu(S)$ is

$$\exp\left(-\lambda \int [1 - E\{e^{-\theta\mu(S+\boldsymbol{t})}\}]d\boldsymbol{t}\right), \qquad \theta \geqq 0.$$

The two examples mentioned by Pyke are now easily treated. If $m = 2$, S is the unit disc and μ is Lebesgue measure over random line segments, $\nu(S)$ measures the total length of lines contained in the unit disc. Clearly, $\sup_{\boldsymbol{t}} \mu(S+\boldsymbol{t}) \leqq 2$, so from the theorem $\nu(S)$ is finite iff the mean length of a line is finite. An exactly similar conclusion holds if μ is $N(0,\sigma^2)$ along a randomly chosen direction, with σ^2 possibly random.

Of course, the problem of determining the distribution of $\nu(S)$ in either of these cases is quite another matter, and will be very messy algebraically; even the integral expression for the mean of $\nu(S)$ is exceedingly complicated.

REFERENCES

BARTLETT, M. S. (1966). *An Introduction to Stochastic Processes*, 2nd. ed. Cambridge University Press.

COX, D. R. & MILLER, H. D. (1968). *The Theory ef Stohastic Processes*. London: Methuen.

DALEY, D. J. (1971). The definition of a multidimensional generalization of shot noise. *J. Appl. Prob.* **8,** 128–35.

DALEY, D. J. & VERE-JONES, D. (1972). A summary of the theory of point processes. In *Stochastic Point Processes; Statistical Analysis, Theory and Applications*, ed. P. A. W. Lewis, 299–383. New York: Wiley.

FELLER, W. (1971). *An Introduction to Probability Theory and its Applications*, vol. II., 2nd ed. New York: Wiley.

HARDY, G. H., LITTLEWOOD, J. E. & POLYA, G. (1934). *Inequalities*. Cambridge University Press.

IBRAGIMOV, I. A. & LINNIK, Yu. V. (1971). *Independent and Stationary Sequences of Random Variables*. Groningen: Wolters-Noordhoff.

ITÔ, K. (1972). Poisson point processes attached to Markov processes. *Proc. 6th Berkeley Symp.*, vol. 3, 225–39.

LOÈVE, M. (1963). *Probability Theory*, 3nd ed. New York: Van Nostrand.

SMITH, W. L. (1958). Renewal theory and its ramifications. *J. R. Statist. Soc.* B **20,** 242–302.

TITCHMARSH, E. C. (1939). *The Theory of Functions*, 2nd. ed. Oxford University Press.

WESTCOTT, M. (1971). On existence and mixing results for cluster point processes. *J. R. Statist. Soc.* B **33,** 290–300.

WESTCOTT, M. (1972). The probability generating functional. *J. Aust. Math. Soc.* **14,** 448–66.

A Sequential Procedure for Diffusion Processes

B. M. BROWN
University of Cambridge

1. Introduction

The aim of this paper is to present a sequential procedure for unbiased testing of the hypothesis that a diffusion process $\{X(t), t \geqq 0\}$ has a given drift function $\mu(x)$ and diffusion function $\sigma^2(x)$. The functions μ and σ^2 are assumed to be suitably well-behaved, and the procedure is based upon observing the process until its deviation from the initial position becomes too large, i.e. upon observing $\{X(t), 0 \leqq t \leqq T\}$, where

$$T = \inf(t > 0 \mid X(t) - X(0) = g \text{ or } -h), \tag{1}$$

with g, h being positive constants. Diffusion processes serve as models for many physical and biological phenomena for which use of a stopping time such as T may well be appropriate.

The work was motivated by the desire to find a sequential procedure which would be suitable for use when the diffusion process $\{X(t)\}$ is replaced by a process $\{X_N(t)\}$, whose distributions only approximate to those of X. The use of such an X_N is called a *diffusion approximation*, justified strictly when N is large and X_N is a member of a sequence of processes $\{X_N\}$ with $X_N \to_D X$ as $N \to \infty$, $\to_D$ denoting the convergence in distribution of stochastic processes, i.e. the weak convergence of corresponding probability measures on some function space. (In the case when X is a diffusion, this function space can be a space of continuous functions, since sample paths of a diffusion process are almost surely (a.s.) continuous.)

The power of weak convergence results lies in the fact that weak convergence is preserved by a.s. continuous mappings (Billingsley, 1968, Theorem 5.1), i.e. if $X_N \to_D X$ and f is an X-a.s. continuous functional, then $f(X_N) \to_D f(X)$. Thus a diffusion approximation entails an allied approximation for many functionals on the approximating process. However, in choosing a functional f upon which to base inference procedures concerning the structure of X_N, it should be ensured that f has the above continuity property, needed to make the approximation valid. This is the main difficulty in using weak-convergence-based approximations for statistical procedures on stochastic processes, and loosely speaking, it means that f should not depend too much on the very fine structure of X. The procedure given here uses an f which depends on the underlying process in a very simple way, and indeed has the required continuity property. Hence it is suitable for use with a diffusion approximation. There is another natural restriction on the practical choice of f, and this is discussed in §4.

We note that the general area of applied probability abounds with so-called diffusion approximations, although the formal weak convergence results which should underlie these have not been proved in some cases. For weak convergence results using various Markov processes, see for example Iglehart (1965), Schach (1971), Stone (1963); the book by Billingsley (1968) contains many examples of processes based on sums of independent random variables (r.vs), and sometimes dependent r.vs, converging weakly to Brownian motion.

It seems almost certain that valid weak convergence results will hold for many processes based upon Markov processes or Markov chains, and indeed the fact that these may not yet be formally proven has not inhibited workers from using the appropriate allied diffusion approximation (for example, in epidemiology McNeil, 1972).

Broadly, a diffusion process is a process with the strong Markov property, continuous sample paths, and stationary transition probabilities which are suitably smooth. If a sequence of approximating processes mimics these properties in the limit, we may expect the corresponding weak convergence result to hold.

Our method may be briefly outlined as follows. The hypothesized drift and diffusion functions μ, σ^2 are used to transform the process X to a process which is on natural scale (i.e. is driftless) and thence in turn, by a deformation of the time scale, to a Brownian motion process. The nature of the stopping time T (see (1)) is preserved by these transformations, and our problem thus reduces to unbiased testing for zero drift of a Brownian motion $\{W(t), t \geqq 0\}$, on the basis of observing $\{W(t), 0 \leqq t \leqq T'\}$, where T' is of the same type as T.

This simple problem is treated separately in §3, which is quite self-contained. §2 contains some preliminary material on diffusion processes, and associated computations. The procedure is summarized in §4, where some brief remarks are also made.

From now on we will write "Brownian motion" simply as "B.M.".

2. DIFFUSION PROCESSES

The basic material in this section concerning diffusions, and drift, diffusion and scale functions is found in Breiman (1968, Chapter 16).

Let $\{X(t), t \geqq 0\}$ be a diffusion. Then $X(\cdot)$ has the strong Markov property, continuous sample paths, stationary transition probabilities, and drift function $\mu(\cdot)$ and diffusion function $\sigma^2(\cdot)$ given by

$$\mu(x) = \lim_{h\to 0} h^{-1} \int_{[|X(h)-x|<\varepsilon]} (X(h)-x)dP_x,$$

$$\sigma^2(x) = \lim_{h\to 0} h^{-1} \int_{[|X(h)-x|<\varepsilon]} (X(h)-x)^2 dP_x,$$

for all $\varepsilon > 0$, where P_x is the probability measure for X, conditional on $X(0) = x$. We will assume that $\mu(\cdot)$, $\sigma^2(\cdot)$ are continuous functions, that $\sigma^2(x) > 0$ for all x in the range of X, and that $X(0) = 0$ a.s.. X is said to be on natural scale if $\mu(x) = 0$ for all x. There is a strictly increasing function $u(\cdot)$, called the scale function, such that $\{u(X(t)), t \geqq 0\}$ is a diffusion on natural scale. $u(\cdot)$ is given by

$$u'(x) = du(x)/dx = \exp\left(-\int_{x_0}^{x} 2\mu(z)(\sigma^2(z))^{-1}dz\right), \tag{2}$$

where x_0 is fixed but arbitrary, and the diffusion function of $\{u(X(t)\}$ is

$$\sigma_0^2(y) = \sigma^2(x)(u'(x))^2, \qquad \text{with } y = u(x). \tag{3}$$

But diffusion processes on natural scale can be transformed into B.M. processes by means of a time-deformation. Specifically, let $\{Y(t), t \geqq 0\}$ be a diffusion process on natural scale with diffusion function $\sigma_1^2(y)$. If $\{W(t)\}$ is a B.M. process with unit variance per unit time, we say that $\{t, W(t)\}$ is a B.M. path for $t > 0$.

Let

$$\tau(t) = \int_0^t \sigma_1^2(Y(y))dy.$$

Then $\{\tau(t), Y(t)\}$ is a B.M. path for $t \geqq 0$. We call $\tau(t)$ the *Brownian time* of $Y(t)$, since $\tau(\cdot)$ tells us at what rate time should be accumulated in order to turn $\{Y(t)\}$ into B.M..

Now let $T = \inf(t > 0 | X(t) = g \text{ or } -h)$, where g, h are positive constants. Then

$$T = \inf(t > 0 \mid u(X(t)) = u(g) \text{ or } u(-h)).$$

$\{u(X(t))\}$ is a diffusion on natural scale and has Brownian time

$$\tau(t) = \int_0^t \sigma_0^2(u(X(y)))\,dy$$

$$= \int_0^t \sigma^2(X(y))(u'(X(y)))^2 dy, \quad \text{from (3)},$$

and since $\{\tau(t), u(X(t))\}$ is a B.M. path, it follows that

$$\tau(T) = \int_0^T \sigma^2(X(y))(u'(X(y)))^2 dy \tag{3a}$$

is the first time a B.M., starting from $u(X(0)) = u(0)$ at time $t = 0$, strikes either of the boundaries $u(g)$ or $u(-h)$.

The observed value of $\tau(T)$ can be computed from $\{X(t),\, 0 \leqq t \leqq T\}$ with the aid of (2) (where it will be usual to choose $x_0 = X(0) = 0$). Then a test of the hypothesis that the true drift and diffusion functions are in fact $\mu(\cdot)$ and $\sigma^2(\cdot)$ can be made from the observed value of $\tau(T)$ and the knowledge of $u(g)$, $u(-h)$, and which of these boundaries is struck at time T, by using the method described in §3.

3. A sequential test of Brownian motion drift

Our aim in this section is to obtain a simple sequential procedure for unbiased testing of the hypothesis $\mu = 0$ against a general alternative, where μ is the unknown drift of a B.M. process $\{W(t), t \geqq 0\}$, with $W(0) = 0$ a.s.. We note that observing a B.M. process through any positive length of time determines the variance of the process (for example from Lévy's result

$$\lim_{n\to\infty} \sum_{j=1}^{2^n} (W(jt2^{-n}) - W((j-1)t2^{-n}))^2 = \text{var } W(t) \text{ a.s.}),$$

so we assume var $W(t)$ to be known and $= t$, for all $t \geqq 0$. We suppose that observation of the process must be terminated when the fluctuation from its initial position becomes too large.

This problem may be restated as follows. Let $EW(t) = \mu t$ and var $W(t) = t$ for all t, where μ is unknown, and let $T = \inf(t > 0 | W(t) = g \text{ or } -h)$, where g, h are positive constants. Given $\{W(t), 0 \leqq t \leqq T\}$, how can an unbiased test of the hypothesis of zero drift, $\mu = 0$, be made?

We propose to use a critical region of the form

$$[T \leqq a,\, W(T) = -h] \cup [T \leqq b,\, W(T) = g],$$

where a, b are to be chosen. This kind of test suggests itself on the following intuitive grounds. If W has zero drift, we expect a relatively long time to pass before the fluctuations of W force it to strike either of the boundaries g or $-h$. On the other hand, if $\mu \neq 0$, there is a steady drift towards either the upper or lower boundary, and we expect a shorter time to elapse before a boundary is reached.

The problem then remains of finding the constants a and b, which evidently must satisfy the following two requirements.

(i) The test has fixed size $\varepsilon > 0$.

(ii) The test is unbiased, i.e. the power function, as a function of μ, has minimum value ε at $\mu = 0$.

At this point, some notation must be introduced. Let

$$P_+(t,\mu) = P(T \leqq t,\, W(T) = g \,|\, \mu),$$

$$P_-(t,\mu) = P(T \leqq t,\, W(T) = -h \,|\, \mu),$$

$$f(t,x,\mu) = \frac{\partial}{\partial x} P(T > t,\, W(t) \leqq x \,|\, \mu),$$

and for $r = 0,1,2$

$$I_r(t,\mu) = \int_{-h}^{g} x^r f(t,x,\mu)dx.$$

We also write $P_+(t,0) = P_+(t)$, $P_-(t,0) = P_-(t)$, $f(t,x,0) = f(t,x)$ and $I_r(t,0) = I_r(t)$, $r = 0,1,2$. The requirements (i), (ii) now become

$$\varepsilon = P_-(a) + P_+(b) \tag{4}$$

and

$$\left[\frac{\partial}{\partial \mu}(P_-(a,\mu) + P_+(b,\mu))\right]_{\mu=0} = 0. \tag{5}$$

The probabilities P_+, P_- are the distribution functions of T on the upper and lower boundaries, respectively; f is the density function of W before the boundaries are reached, and the I_r are moments associated with f. The Laplace transforms of P_+, P_- are found by Breiman (1968, page 289) but are very difficult to invert. We will try an alternative approach to obtain information about P_+, P_- by using the densities f.

Firstly, we note that

$$I_0(t,\mu) = 1 - P_+(t,\mu) - P_-(t,\mu). \tag{6}$$

Next, for fixed $t > 0$ let $T(t) = \min(T,t)$. Then $T(t)$ is a stopping time and $\{W(T(t)) - \mu T(t),\ t \geqq 0\}$ is a martingale with zero mean for all $t \geqq 0$, i.e.

$$\begin{aligned} &gP_+(t,\mu) - hP_-(t,\mu) + I_1(t,\mu) \\ &= \mu \int_0^t y\,d(1 - I_0(y,\mu)) + \mu t I_0(t,\mu) \\ &= \mu \int_0^t I_0(y,\mu)dy. \end{aligned} \tag{7}$$

Also, $T(t) \leqq t$, so that $ET(t) < \infty$, and the second order Wald equation $E(W(T(t)) - \mu T(t))^2 = ET(t)$ holds (see for example Hall, 1970), i.e. for $\mu = 0$,

$$g^2P_+(t) + h^2P_-(t) + I_2(t) = \int_0^t yd(1 - I_0(y)) + tI_0(t)$$

$$= \int_0^t I_0(y)dy. \tag{8}$$

The last three equations suggest that (4) and (5) can be rewritten in terms of the I_r, $r = 0, 1, 2$, which depend in turn on the densities f. We therefore now examine f. Let $\phi(u)$ $(2\pi)^{-\frac{1}{2}}\exp(-u^2/2)$, the standard normal density function. When $\mu = 0$,

$$f(t, x) = t^{-\frac{1}{2}} \sum_{j=-\infty}^{\infty} (\phi(t^{-\frac{1}{2}}(x + 2j(g + h))) - \phi(t^{-\frac{1}{2}}(x - 2g + 2j(g + h))). \tag{9}$$

This formula is obtainable directly by multiple use of the reflection principle; see Feller (1966, page 329). For $\mu \neq 0$, there is a similar expression, namely

$$f(t, x, \mu) = \exp(-\mu^2 t/2 + \mu x) \cdot f(t, x). \tag{10}$$

The formal verification of (10) also depends upon the reflection principle, but the details are quite tedious and are omitted. However, the underlying idea is simple. Given

$$W(t) = y \in [-h, g] \text{ for some fixed } y \text{ then } P(T > t \mid \mu, W(t) = y)$$

no longer depends upon μ and the reflection principle can be used as in the development of (9). Averaging over $W(t) = y \in [-h, g]$ then forces μ to reappear as on the right-hand side of (10).

Equation (10) can now be used, for small μ, to evaluate the partial derivatives at $\mu = 0$ which occur in (5). From (6) and (7) we have

$$(g + h)P_+(t, \mu) = h - hI_0(t, \mu) + \mu \int_0^t I_0(y, \mu)dy - I_1(t, \mu) \tag{11}$$

and

$$(g + h)P_-(t, \mu) = g - gI_0(t, \mu) - \mu \int_0^t I_0(y, \mu)dy + I_1(t, \mu). \tag{12}$$

But (10) tells us that $\lim_{\mu \to 0} \mu^{-1}(f(t, x))^{-1}(f(t, x, \mu) - f(t, x)) = x$ uniformly for bounded t and $-h \leqq x \leqq g$, so that $(\partial/\partial\mu)I_r(t, \mu) = I_{r+1}(t, \mu)$, $r = 0, 1$. Hence from (11) and (12) we obtain

$$(g + h)\left[\frac{\partial}{\partial\mu} P_+(t, \mu)\right]_{\mu=0} = -hI_1(t) + \int_0^t I_0(y)dy - I_2(t) \tag{13}$$

and

$$(g + h)\left[\frac{\partial}{\partial\mu} P_-(t, \mu)\right]_{\mu=0} = -gI_1(t) - \int_0^t I_0(y)dy + I_2(t). \tag{14}$$

Finally, systematic use of (8), (11), (12), (13) and (14) enable (4) and (5) to be rewritten simply as

$$(1 - \varepsilon) = I_0(a) - g^{-1}I_1(a)$$

and

$$(1 - \varepsilon) = I_0(b) + h^{-1}I_1(b). \tag{15}$$

Values of the functions I_0, I_1 can be found with good accuracy by integrating in (9) and then using tables of the exponential function and the standard normal distribution function. Hence, solutions for (a, b) are readily obtainable from (15). Only the leading terms of (9) make a non-negligible contribution, and for most purposes the resulting approximations to (15),

$$g\varepsilon/(2(g+h)) = \Phi(a^{-\frac{1}{2}}(2g+h)) - \Phi(ha^{-\frac{1}{2}})$$

and (16)

$$h\varepsilon/(2(g+h)) = \Phi(b^{-\frac{1}{2}}(g+2h)) - \Phi(gb^{-\frac{1}{2}}),$$

will be adequate, where Φ is the standard normal distribution function. We note that a and b are of the form $(g+h)^2\beta(g/(g+h))$, i.e. are proportional to $(g+h)^2$. Hence, to determine (a, b) for all pairs (g, h), with ε fixed, it suffices to find the values of (a, b) corresponding to $g+h=1$, $h=\lambda$, $g=1-\lambda$, for $0<\lambda<1$. If this is done, then it will be enough to find only the values of a, for if we write $a=a(\lambda)$, $b=b(\lambda)$, then the relation $b(\lambda)=a(1-\lambda)$ yields values of b. A table of illustrative values is given below for $g+h=1$ and $\varepsilon=.1$, .05, and .01.

Finally, it is easily shown from (11), (12), and (15) that

$$\varepsilon = P_-(a)/P_-(\infty)$$

and

$$\varepsilon = P_+(b)/P_+(\infty),$$

thus showing that the test considered here (unbiased, of size ε) is exactly the same as a natural size ε conditional test, which, when given $T=t$ with absorption on the upper boundary, uses critical regions of the form $[T \leqq b]$, and computes probabilities conditional on the information that absorption was at the upper boundary before the lower (with a similar test for absorption at the lower boundary). This fact is rather appealing. Perhaps there is a way of looking at the problem which makes it intuitively obvious, but if there is, it has escaped the author.

Table. *Values of* a *for which the critical region* $[T \leqq a, W(T) = -\lambda] \cup [T \leqq b, W(T) = 1-\lambda]$ *describes an unbiased size* ε *test of the hypothesis that* $\mu = 0$, *against a general alternative. The relation* $b(\lambda) = a(1-\lambda)$ *yields values of* b.

λ \ ε	.10	.05	.01
.1	.00348	.00249	.00147
.2	.0131	.00948	.00569
.3	.0274	.0202	.0124
.4	.0452	.0340	.0212
.5	.0651	.0498	.0317
.6	.0854	.0665	.0435
.7	.104	.0828	.0556
.8	.119	.0967	.0670
.9	.127	.105	.0765

4. Discussion

Our procedure for unbiased testing of the hypothesis that a diffusion process has a prescribed drift function $\mu(\cdot)$ and a prescribed diffusion function $\sigma^2(\cdot)$ may be summarized as follows.

Observe the process $\{X\}$ up to the stopping time T. Also, compute u', then u, from the hypothesized $\mu(\cdot)$, $\sigma^2(\cdot)$, using (2). Then, $u'(\cdot)$, $\sigma^2(\cdot)$ and $\{X(t), 0 \leqq t \leqq T\}$ determine $\tau(T)$, using (3a). $\tau(T)$ is the first time a B.M. starting from $u(X(0)) = u(0)$ strikes either $u(g)$ (if $X(T) = g$) or $u(-h)$ (if $X(T) = -h$). Values of a, b which determine the critical region

$$[\tau(T) \leqq a, X(T) = -h] \cup [\tau(T) \leqq b, X(T) = g]$$

are found by the method of §3, for a test of prescribed size ε. Alternatively, the approximate significance level of the realized value of $\tau(T)$ is given by (cf. (16))

$$2(u(g)-u(-h))(u(g)-u(0))^{-1}\{\Phi((\tau(T))^{-\frac{1}{2}}(2u(g)-u(0)-u(-h)))-\Phi((\tau(T))^{-\frac{1}{2}}(u(0)-u(-h)))\}$$

when $X(T) = -h$, or

$$2(u(g)-u(-h))(u(0)-u(-h))^{-1}\{\Phi((\tau(T))^{-\frac{1}{2}}(u(g)+u(0)-2u(-h)))-\Phi((\tau(T))^{-\frac{1}{2}}(u(g)-u(0)))\}$$

when $X(T) = g$.

It would be most desirable to be able to find unbiased confidence regions of prescribed confidence levels for the unknown drift and diffusion functions $\mu(\cdot)$, $\sigma^2(\cdot)$ of a diffusion process (or diffusion approximation), and in principle, our procedure is able to do this. Formally, the confidence region for $(\mu(\cdot), \sigma^2(\cdot))$ consists of those points which are not rejected by a test, of prescribed size, of the form described above (see Lehmann, 1959, page 176). In practice, however, this fact alone is of little use, for it is not yet known what shape the resulting confidence regions have, or even whether they are simply connected. This means that *all* points $(\mu(\cdot), \sigma^2(\cdot))$ (or rather, a great many of them) must be tested to see whether they are eligible for the confidence region, and the task is clearly impractical. Even in the simplest case, when the unknown drift μ of a B.M. is to be investigated, it is not clear that the confidence region obtained will be an interval (though all examples tried suggest that it must be).

This special case perhaps deserves one further remark. In looking for confidence intervals for μ by the method described, one meets the moment generating function of the sample local time function of a B.M.. Local time (see Itô & McKean, 1965, Chapter 2) is one of the more difficult parts of the theory of diffusion processes, and the fact that it appears in the context of the present seemingly simple problem is interesting, but rather daunting.

Finally, we note that, in practice, the process $\{X(t)\}$ (or the diffusion approximation $\{X_N(t)\}$ will not be observed throughout continuous time, but rather, on a discrete skeleton of time points. The method described in this paper will be suitable for use only when these time points are close together, so as to mimic the continuous-time observations necessary to find the stopping time T and the Brownian time $\tau(T)$. In addition, the method depends on the assumption of §3, that continuous time observations are available, thus determining the variance of B.M.. For observations over more widely spaced points of time, quite different methods will be appropriate. In terms of the discussion in §1, these methods will employ a functional f which depends on X only through its values over a discrete set of time-points.

REFERENCES

BILLINGSLEY, P. (1968). *Convergence of Probability Measures*. New York: Wiley.

BREIMAN, L. (1968). *Probability*. Reading, Mass.: Addison-Wesley.

FELLER, W. (1966). *An Introduction to Probability Theory and its Applications*. II. New York: Wiley.

HALL, W. J. (1970). On Wald equations in continuous time. *J. Appl. Prob.* **7**, 59–68.

IGLEHART, D. L. (1965). Limit diffusion approximations for the many server queue and the repairman problem. *J. Appl. Prob.* **2**, 429–41.

ITÔ, K. & MCKEAN, H. P. (1965). *Diffusion Processes and their Sample Paths*. Berlin: Springer-Verlag.

LEHMANN, E. L. (1959). *Testing Statistical Hypotheses*. New York: Wiley.

MCNEIL, D. R. (1972). On the simple stochastic epidemic. *Biometrika* **59**, 494–7.

SCHACH, S. (1971). Weak convergence results for a class of multivariate Markov processes. *Ann. Math. Statist.* **42**, 451–65.

STONE, C. (1963). Limit theorems for random walks, birth and death processes, and diffusion processes. *Ill. J. Math.* **7**, 638–60.

Distribution-Free Confidence Intervals Based on Linear Rank Statistics*

G. E. NOETHER
The University of Connecticut

1. Introduction

It is probably no exaggeration to say that at least 90% of all contributions to nonparametric inference are concerned with hypothesis testing. Nonparametric estimation in general and distribution-free confidence intervals in particular seem to have attracted much less attention. In the parametric case the close connection between hypothesis testing and confidence intervals is brought out in even the most elementary statistics texts. In the nonparametric field, there still seems to prevail a rather widepread belief that with few exceptions nonparametric tests do not lend themselves to conversion into confidence intervals except possibly by trial and error methods.

In this essay we describe and illustrate a systematic procedure for converting nonparametric tests of randomness based on linear rank statistics into confidence intervals for suitably defined parameters. Our approach unifies and generalizes work by Bauer (1972) and Walter (1962).

We do not concern ourselves with the interesting question of which particular rank procedure should be used in any particular case. This problem is treated in great detail for the case of hypothesis testing in books by Hájek (1969) and Hájek & Šidák (1967). Investigations by Lehmann (1963) and Hodges & Lehmann (1963) suggest that estimation procedures based on the various tests show corresponding behaviour.

It is well-known that tests based on linear rank statistics are distribution-free under random sampling from continuous populations. For the time being, we shall make this assumption to avoid unnecessary complications. As we shall see in Section 6, problems arising from ties among the observations can be resolved much more easily for confidence intervals than for hypothesis testing.

2. A general method for finding confidence intervals

We consider tests of the hypothesis $\theta = \theta_0$ against the alternative $\theta \neq \theta_0$ of the following type. We have computed G quantities $u_1 \leqq u_2 \leqq \cdots \leqq u_G$ from the sample observations (for continuous populations, all u's will be different with probability 1). Associated with u_g, $g = 1, \cdots, G$, is a nonnegative quantity d_g, such that the null hypothesis is rejected if either

* This article has been written with support from the National Science Foundation under Grant GP–11470.

$$T_- = \sum_{u_g < \theta_0} d_g \leqq c_1$$

or (1)

$$T_+ = \sum_{u_g > \theta_0} d_g \leqq c_2 .$$

To convert a test of type (1) into a confidence interval for θ, we note that $T_- = T_-(\theta)$ is an increasing step function of θ, while $T_+ = T_+(\theta)$ is a decreasing step function of θ, with steps occurring at u_g's for which $d_g > 0$. Let θ^* and θ^{**} be the confidence limits of a confidence interval,

$$\theta^* < \theta < \theta^{**} . \tag{2}$$

Since the interval (2) consists of θ-values for which both $T_-(\theta) > c_1$ and $T_+(\theta) > c_2$, we have

$$\theta^* = u_{g^*} \text{ and } \theta^{**} = u_{G+1-g^{**}}, \tag{3}$$

where g^* is the smallest integer such that

$$d_1 + d_2 + \cdots + d_{g^*} > c_1$$

and g^{**} is the smallest integer such that

$$d_G + d_{G-1} + \cdots + d_{G+1-g^{**}} > c_2 .$$

The determination of θ^* and θ^{**} is particularly simple when all d_g equal either 0 or some positive constant d. In this case, let h_i be the smallest integer such that $h_i d > c_i$, $i = 1,2$. Then θ^* and θ^{**} are the h_1th smallest and h_2th largest of the u's with $d_g = d$.

The following example illustrates the approach. Let η be the median of a population from which we have a random sample $z_1, \cdots, z_N$. The hypothesis $\eta = \eta_0$ can be tested by means of the sign test, which rejects the null hypothesis if either $T_- = \#(z_k < \eta_0)$ or $T_+ = \#(z_k > \eta_0)$ is $\leqq c$. In this form, the sign test is of type (1) with u_g being the gth order statistic of the sample and $d_g = 1$. It follows that $g^* = g^{**} = c + 1$, and the confidence interval (2) is the usual confidence interval for the median bounded by the $(c + 1)$th smallest and largest observations in the sample.

The linear rank tests to be considered in Sections 3, 4 and 5 are usually based on a single test statistic S. If S varies between $s_{\min}$ and $s_{\max}$, we may set

$$T_- = S - s_{\min}$$

and (4)

$$T_+ = s_{\max} - S ,$$

so that $c_1 = s_1 - s_{\min}$ and $c_2 = s_{\max} - s_2$, where s_1 and s_2 are the lower and upper critical values for the two-sided test based on S. In particular, if the distribution of S is symmetric, T_- and T_+ have identical distributions and $c_1 = c_2 = c$.

In Section 3 we discuss confidence intervals for the centre of a symmetric distribution. In Section 4, we take up the case of linear regression, and in Section 5 we specialize the regression model to the two-sample shift-in-location model.

3. ONE-SAMPLE CASE

Let η be the centre of a symmetric population. We want to find a confidence interval for η based on a random sample $z_1, \cdots, z_N$, which, for notational simplicity, we shall assume to be arranged according to size. Let

$$0 = a_0 \leqq a_1 \leqq \cdots \leqq a_N \tag{5}$$

be a set of scores. Let $r[v_h]$ denote the rank of v_h among quantities $v_1, \cdots, v_N$. If $r[|z_h - \eta_0|] = k$, we define

$$t_k(\eta_0) = t_k = \begin{matrix}1\\0\end{matrix} \text{ depending on whether } z_h \begin{matrix}<\\>\end{matrix} \eta_0 .$$

According to Hájek (1969), the linear rank statistic corresponding to the scores (5) for testing the hypothesis $\eta = \eta_0$ against the alternative $\eta \neq \eta_0$ may be written as

$$S = \sum_{k=1}^{N} t_k a_k .$$

We easily find $s_{\min} = 0$ and $s_{\max} = a_1 + \cdots + a_N$, so that (4) becomes

$$T_- = S = \sum t_k a_k$$

and (6)

$$T_+ = \sum (1 - t_k) a_k .$$

(The sign test is of this form with scores $a_1 = \cdots = a_N = 1$.) Since S is symmetrically distributed, $c_1 = c_2 = c$, where c is the lower $\alpha/2$-level critical value of S.

We now show that (6) is of type (1) and determine the appropriate u's and d's. T_- is an increasing function of η that may change value only if two successive scores a_r and a_{r+1} exchange places in contributing to T_- and T_+, respectively. This happens if either η passes from the left to the right of an observation z_i or from the left to the right of a midpoint of two observations z_i and z_j. Thus the u's are the $\frac{1}{2}N(N+1)$ averages

$$u_{ij} = \tfrac{1}{2}(z_i + z_j), \quad 1 \leqq i \leqq j \leqq N. \tag{7}$$

Since there are $(j-i-1)$ observations between z_i and z_j, the ranks associated with z_i and z_j are $(j-i)$ and $(j-i+1)$, respectively. (This holds even when $i = j$.) Thus

$$d_{ij} = a_{j-i+1} - a_{j-i} .$$

For the Wilcoxon signed rank test we have $a_k = k$, $k = 1, \cdots, N$, so that $d_{ij} = 1$. It then follows from the results of Section 2 that the confidence interval corresponding to the Wilcoxon signed rank test is bounded by the $(c+1)$th smallest and largest sample averages (7). This is of course a well-known result.

4. Linear regression

(i) *The Model.* We consider observations z_k, $k = 1, \cdots, N$, satisfying the following linear regression model,

$$z_k = \eta - \Delta\tau_k + e_k , \tag{8}$$

where the e_k represent a random sample from a population with median 0 and the τ_k are known constants satisfying

$$\tau_1 \leqq \tau_2 \leqq \cdots \leqq \tau_N .$$

By setting $\tau_1 = \cdots = \tau_m = 0$, $\tau_{m+1} = \cdots = \tau_N = 1$, model (8) reduces to the two-sample shift-in-location model with shift parameter Δ, which will be discussed in Section 5.

(ii) *Confidence intervals for* Δ. According to Hájek (1969), linear rank tests for testing the hypothesis $\Delta = 0$ are based on statistics

$$S = \sum_{k=1}^{N} \tau_k a(R_k), \tag{9}$$

where $R_k = r[z_k]$ and the scores $a(k) = a_k$ satisfy the condition

$$a_1 \leqq \cdots \leqq a_N. \tag{10}$$

We have

$$s_{\min} = \tau_1 a_N + \cdots + \tau_N a_1 \tag{11}$$

and

$$s_{\max} = \tau_1 a_1 + \cdots + \tau_N a_N. \tag{12}$$

The hypothesis $\Delta = \Delta_0$ is tested by replacing the ranks R_k in (9) by

$$R_k(\Delta_0) = r[z_k + \Delta_0 \tau_k].$$

As a consequence, the statistic S, and therefore the statistics T_- and T_+, considered as functions of Δ change value whenever two adjusted observations $z_i + \Delta\tau_i$ and $z_j + \Delta\tau_j$ interchange rank order. This occurs as Δ passes from the left to the right of

$$u_{ij} = \frac{z_i - z_j}{\tau_j - \tau_i}, \quad 1 \leqq i < j \leqq N, \quad \tau_i \neq \tau_j. \tag{13}$$

Let r and $r+1$ be the ranks involved in the interchange of $z_i + \Delta\tau_i$ and $z_j + \Delta\tau_j$. The increase in the value of the statistic T_- then amounts to

$$\begin{aligned} d_{ij} &= \tau_i(a_r - a_{r+1}) + \tau_j(a_{r+1} - a_r) \\ &= (\tau_j - \tau_i)(a_{r+1} - a_r). \end{aligned} \tag{14}$$

The appropriate ranks r and $r+1$ are found by computing and ordering the adjusted observations $z_k + u_{ij}\tau_k$. If the scores are equally spaced, these operations become unnecessary. In particular, if $a_k = k$, we simply have $d_{ij} = \tau_j - \tau_i$. If also $\tau_k = k$, the linear rank test (9) is equivalent to the Spearman rank correlation test and the confidence interval for Δ is determined by

$$u_{ij} = \frac{z_i - z_j}{j - i}, \quad d_{ij} = j - i, \quad 1 \leqq i < j \leqq N.$$

It is of some interest to ask the following question. Suppose we simply use symmetrically located u_{ij}'s in the sequence (13) as our confidence bounds. What is the corresponding test of randomness? Since our procedure implies that $d_{ij} = 1$, we have, for example

$$\begin{aligned} T_+ &= \#(u_{ij} > \Delta_0) = \#[z_i - z_j > \Delta_0(\tau_j - \tau_i)] \\ &= \#(z_i + \Delta_0\tau_i > z_j + \Delta_0\tau_j) = \#(e_i > e_j), \quad 1 \leqq i < j \leqq N. \end{aligned}$$

The appropriate test is Mann's test for testing the hypothesis of randomness against a monotone trend. The test statistic is based on the Kendall rank correlation statistic, which is not a linear rank statistic. Clearly, the approach of Section 2 is not limited to linear rank statistics.

(iii) *Confidence intervals for* η. Before turning to the two-sample case, we briefly look at the problem of finding a confidence interval for the parameter η in model (8). If we set $z'_k = z_k + \Delta\tau_k = \eta + e_k$, the z'-sample constitutes a random sample from a population with median η. Thus for known Δ, depending on whether or not we are willing to assume that the e_k are symmetrically distributed, the methods of Section 3 or the sign test procedure of Section 2 are appropriate. When Δ is unknown, as will usually be the case, we may follow a suggestion of Adichie (1967) to replace Δ by an estimate $\hat{\Delta}$ when computing z'_k. If we have determined a confidence interval for Δ, a simple point estimate is given by

$$\hat{\Delta} = \tfrac{1}{2}(\Delta^* + \Delta^{**}), \tag{15}$$

the midpoint of the confidence interval. Adichie uses an estimate of the same form as (15), except that Δ^* and Δ^{**} are computed by using the medians of the sampling distributions of T_- and T_+ in place of the critical values c_1 and c_2. In particular, the median of the u_{ij}'s given by (13) is a possible estimate $\hat{\Delta}$.

5. The two-sample case

We now consider the special case of model (8) when $\tau_1 = \cdots = \tau_m = 0, \tau_{m+1} = \cdots = \tau_{m+n} = 1$, $m + n = N$. It is convenient to introduce the following special notation. The observations $z_1, \cdots, z_m$ corresponding to $\tau_k = 0$ are rearranged to give $x_1 \leqq \cdots \leqq x_m$; the observations $z_{m+1}, \cdots, z_N$ corresponding to $\tau_k = 1$ are rearranged to give $y_1 \leqq \cdots \leqq y_n$. Model (8) implies that $x_1, \cdots, x_m$ and $y_1, \cdots, y_n$ constitute two (ordered) random samples for two random variables X and Y, such that X and $Y + \Delta$ are identically distributed. We want to apply the results of Section 4(ii) to the two-sample shift parameter Δ.

From (9), the appropriate linear rank statistic becomes

$$S = \sum_{k=m+1}^{N} a(R_k) = \sum_{j=1}^{n} a(r[y_j]), \tag{16}$$

the sum of scores corresponding to y-observations in the overall ranking of both x- and y-observations. From (11) and (12), $s_{\min} = a_1 + \cdots + a_n$ and $s_{\max} = a_{m+1} + \cdots + a_{m+n}$. From (13),

$$u_{ij} = x_i - y_j, \ i = 1, \cdots, m; \ j = 1, \cdots, n \tag{17}$$

and, according to (14),

$$d_{ij} = a_{r+1} - a_r = a_{i+j} - a_{i+j-1},$$

since x_j and $y_i + \Delta$ occupy rank positions $i + j - 1$ and $i + j$ as Δ passes from the left to the right of $u_{ij} = x_i - y_j$.

We consider two special cases. The test statistic (16) reduces to the Wilcoxon two-sample statistic for scores $a_k = k$. The corresponding statistics T_- and T_+ are the Mann-Whitney statistics

$$T_- = \#(y_j > x_i) \text{ and } T_+ = \#(x_i > y_j),$$

and according to Section 2 the confidence interval for Δ is bounded by the $(c + 1)$th smallest and largest difference (17), where c is the lower critical value of the Mann-Whitney statistic.

The median test offers an example where we need not consider all mn differences (17). As test statistic we use

$$S = \#(y_j > \text{combined sample median}). \tag{18}$$

To illustrate the approach we assume that $m+n = N = 2p$. The appropriate scores are then $a_1 = \cdots = a_p = 0$, $a_{p+1} = \cdots = a_N = 1$. If we assume that $n \leqq m$, we have $T_- = S$. We note that $d_{ij} = 0$ unless $i+j = p+1$, in which case $d_{ij} = 1$. Thus the confidence interval is bounded by the $(c+1)$th smallest and largest differences $x_i - y_j$ with $i+j = p+1$, where c is the lower critical value of the median test statistic (18). The case $m+n = 2p+1$ can be handled correspondingly.

6. Samples from discontinuous populations

It is well-known that for samples from discontinuous populations nonparametric procedures are no longer distribution-free. As far as linear rank tests—as well as most other nonparametric tests—are concerned, discontinuity raises the additional question of how to redefine the test statistic, to take care of possible ties among the observations.

Somewhat surprisingly, in general ties do not require any new definitions as far as confidence intervals are concerned. The u's and associated d's are computed according to rule. If two or more u's turn out to be equal, the amount of change in the value of the test statistic at $\theta = u$ equals the sum of the corresponding d's.

As in Noether (1967), the projection method gives the following result concerning the confidence coefficient,

$$\gamma_{\text{op}} \leqq \gamma \leqq \gamma_{\text{cl}},$$

where γ is the confidence coefficient associated with the distribution-free case and γ_{op} and γ_{cl} refer to open and closed intervals.

References

ADICHIE, J. N. (1967). Estimates of regression parameters baseed on rank tests. *Ann. Math. Statist.* **38,** 894–904.

BAUER, D. F. (1972). Constructing confidence sets using rank statistics. *J. Am. Statist. Ass.* **67**, 687–90.

BAUER, D. F. (to appear). On some parametric estimates for regression parameters. *J. Am. Statist. Ass.*

HÁJEK, J. (1969). *Nonparametric Statistics.* San Francisco: Holden-Day.

HÁJEK, J. & ŠIDÁK, Z. (1967). *Theory of Rank Tests.* New York: Academic Press.

HODGES, J. L. & LEHMANN, E. L. (1963). Estimates of location based on rank tests. *Ann. Math. Statist.* **34,** 598–611.

LEHMANN, E. L. (1963). Nonparametric confidence intervals for a shift parameter. *Ann. Math. Statist.* **34,** 1507–12.

NOETHER, G. E. (1967). Wilcoxon confidence intervals for location parameters in the discrete case. *J. Am. Statist. Ass.* **62,** 184–8.

WALTER, E. (1962). Verteilungsunabhängige Schätzverfahren. *Z. Angew. Math. Mech.* **42,** T85–7.

Sequential Estimation of p in Bernoulli Trials*

HERBERT ROBBINS AND DAVID SIEGMUND
Columbia University

1. Sequential estimation of p in Bernoulli trials

Let $x_1, \cdots$ be the outcomes of a sequence of Bernoulli trials with $P(x_i = 1) = p$, $P(x_i = 0) = q = 1 - p$, the parameter $0 < p < 1$ being unknown. In many cases we wish to estimate p, or a function of p, subject to some condition which necessitates using a random sample size N.

Example 1. To estimate p in such a way that the expected value of $\{(\text{estimate} - p)^2/(p^2q^2)\}$ shall be $\approx 1/c$, where c is some (large) preassigned constant. If we use s_n/n as our estimate of p, where $s_n = x_1 + \cdots + x_n$, then for any fixed number of observations

$$\frac{E(s_n/n - p)^2}{p^2q^2} = \frac{1}{npq} \approx \frac{1}{c}$$

only if $npq \approx c$, which requires a knowledge of p. Since, however, for large n

$$s_n/n \approx p, \ (n - s_n)/n \approx q,$$

making $npq \approx c$ is more or less the same as making $n \cdot (s_n/n) \cdot ((n - s_n)/n) \approx c$. It is therefore natural to consider the random sample size N defined by

$$N = \text{first } n \geqq 1 \text{ such that } s_n(n - s_n) > cn, \tag{1}$$

and to inquire whether

$$\frac{cE(s_N/N - p)^2}{p^2q^2} \to 1 \quad \text{as } c \to \infty \tag{2}$$

for fixed $0 < p < 1$. We shall see that this is so.

Example 2. To estimate $\log(p/q)$ in such a way that the estimate shall be approximately normally distributed with mean $\log(p/q)$ and preassigned variance $1/c$. Since for large n the estimate

$$\log\left(\frac{s_n + \frac{1}{2}}{n - s_n + \frac{1}{2}}\right)$$

(proposed by Haldane, 1956) is approximately normal with mean $\log(p/q)$ and variance $\approx 1/npq$, we can hope to make the latter quantity $\approx 1/c$ by again using the random sample size N defined by (1). We shall see that in fact

* Research supported by National Science Foundation grant no. NSF–GP–33570X.

$$u \equiv c^{\frac{1}{2}}\left\{\log\left(\frac{s_N+\frac{1}{2}}{N-s_N+\frac{1}{2}}\right)-\log(p/q)\right\} \to N(0,1) \tag{3}$$

in distribution as $c \to \infty$ for any fixed $0 < p < 1$. This enables us to give confidence intervals for $\log(p/q)$ of preassigned width and approximately known coverage probability, when c is large.

We begin by observing that since $s_n(n-s_n) \leqq n^2/4$, from (1) it follows that necessarily

(a) $$N > 4c.$$

In the following section we shall show that

(b) $$pqN/c \to 1 \text{ in probability as } c \to \infty,$$

and also

(c) $$pq\,EN/c \to 1 \quad \text{as } c \to \infty.$$

Using (a), (b), and (c) we shall now establish (2) and (3).

Proof of (2). We have

$$\begin{aligned}\frac{c(s_N/N-p)^2}{p^2q^2} &= \frac{(s_N-Np)^2}{c}+\frac{(s_N-Np)^2}{c}\left(\frac{c^2}{N^2p^2q^2}-1\right)\\ &\equiv I_1+I_2, \text{ say.}\end{aligned}$$

Now by (a),

$$-1 \leqq \frac{c^2}{N^2p^2q^2}-1 \leqq \frac{c^2}{N^2p^2q^2} \leqq \frac{1}{16p^2q^2}, \tag{4}$$

and by (b),

$$\frac{c^2}{N^2p^2q^2}-1 \to 0 \text{ in probability as } c \to \infty. \tag{5}$$

Also, by (b) and the Doeblin-Anscombe theorem (see e.g. Chung, 1968, p. 197)

$$w \equiv \frac{s_N-Np}{(Npq)^{\frac{1}{2}}} \to N(0,1) \text{ in distribution as } c \to \infty, \tag{6}$$

and hence again by (b)

$$I_1 = \frac{(s_N-Np)^2}{c} \to \{N(0,1)\}^2 \text{ in distribution as } c \to \infty. \tag{7}$$

Moreover, by a generalization of Wald's lemma (Chow, Robbins & Teicher, 1965), as $c \to \infty$

$$EI_1 = \frac{pqEN}{c} \to 1 \quad \text{by (c).} \tag{8}$$

It follows from (7) and (8) (see e.g. Loève, 1963, p. 183) that the family I_1 of random variables is uniformly integrable in the positive parameter c, and from this and (4) it follows that the family I_2 is also. Hence, from (5), $EI_2 \to 0$ as $c \to \infty$, so from (8)

$$\frac{cE(s_N/N - p)^2}{p^2q^2} = EI_1 + EI_2 \to 1,$$

which proves (2).

Proof of (3). Referring to (6), we see by algebra that for any fixed t

$$P(u \leqq t) = P\left(w \leqq \frac{N^{\frac{1}{2}}pq(e^{tc^{-\frac{1}{2}}} - 1) + (pe^{tc^{-\frac{1}{2}}} - q)/(2N^{\frac{1}{2}})}{(pq)^{\frac{1}{2}}(q + pe^{tc^{-\frac{1}{2}}})}\right).$$

Since $e^{tc^{-\frac{1}{2}}} = 1 + tc^{-\frac{1}{2}} + o(c^{-\frac{1}{2}})$ and $N \sim c/pq$ in probability as $c \to \infty$, the bound on w above is $\sim$ in probability to

$$\frac{N^{\frac{1}{2}}pqt}{(cpq)^{\frac{1}{2}}} \to t \text{ in probability.}$$

Hence (3) follows from (6).

2. The asymptotic distribution of the sample size N.

We begin by proving (b) and (c) of the previous section. We shall in fact obtain strict inequalities for EN((13) below) from which (c) follows.

By (1) we have that

$$s_{N-1}(N - 1 - s_{N-1}) \leqq c(N-1),$$

and by considering separately the two cases $x_N = 1$ and $x_N = 0$ we see that

$$cN < s_N(N - s_N) \leqq (c+1)(N-1)$$

and hence

$$c < s_N(N - s_N)/N < c + 1. \tag{9}$$

Dividing (9) by N and letting $c \to \infty$ (hence $N > 4c \to \infty$), we have by the strong law of large numbers

$$\limsup_{c\to\infty} c/N \leqq pq \leqq \liminf_{c\to\infty} (c+1)/N \text{ a.s.},$$

from which (b) of §1 follows.

It is easy to verify the algebraic identity

$$s_n(n - s_n)/n = (q-p)(s_n - np) + npq - (s_n - np)^2/n. \tag{10}$$

Hence from (9), (10), and (a) of §1, we obtain

$$c < (q-p)(s_N - Np) + Npq < c + 1 + (s_N - Np)^2/(4c). \tag{11}$$

By the generalization of Wald's lemma (Chow, Robbins & Teicher, 1965), if $EN < \infty$, then

$$E(s_N - Np) = 0,\ E(s_N - Np)^2 = pq\,EN, \tag{12}$$

which yield upon substitution into (11)

$$c < pq\,EN < (c+1) + pq\,EN/(4c)$$

and hence

$$c < pq\,EN < (c+1)/(1-(4c)^{-1}). \tag{13}$$

As we remarked before, (c) of §1 follows immediately from (13). (In the course of the preceding derivation we assumed that $EN < \infty$. It is easy to see that this is indeed so by repeating the preceding argument with $\min(N,n)$ in place of N to obtain the right-hand inequality of (13) and then letting $n \to \infty$.)

We now turn to an examination of the asymptotic distribution of N, which leads to surprisingly different results in the cases $p = \frac{1}{2}$ and $p \neq \frac{1}{2}$.

Let $[4c]$ denote the least integer $> 4c$. Now for any $n \geqq [4c]$, if $s_n(n-s_n) > cn$, then also $s_{n+1}(n+1-s_{n+1}) > c(n+1)$, and hence

$$P\{N \leqq n\} = P\{s_n(n-s_n)/n > c\}. \tag{14}$$

Case 1. If $p = q = \frac{1}{2}$, then by (10) and (14) we have for $j = 0, 1, \cdots$, putting $n = [4c]+j$,

$$P\{N \leqq [4c]+j\} = P\left\{\frac{(s_n - n/2)^2}{n/4} < [4c] - 4c + j\right\}.$$

Hence, by the central limit theorem, letting G denote the chi-squared distribution function with one degree of freedom,

$$P\{N - [4c] \leqq j\} - G([4c] - 4c + j) \to 0 \tag{15}$$

uniformly in $j = 0, 1, 2, \cdots$ as $n \to \infty$. Note that according to (15) the family $\{N\}$ does *not* have an asymptotic distribution in the usual sense, since $[4c] - 4c$ continues to oscillate (although boundedly) as $c \to \infty$. Nevertheless (15) serves the same purpose as an asymptotic distribution in the usual sense.

Case 2. If $p \neq q$, say $p < q$, then by (10) and (14) for all $n \geqq [4c]$

$$\begin{aligned} P\{N \leqq n\} &= P\left\{(q-p)(s_n - np) - \frac{(s_n - np)^2}{n} > c - npq\right\} \\ &= P\left\{\frac{(q-p)(s_n - np)}{(npq)^{\frac{1}{2}}} - \frac{(s_n - np)^2}{n^{\frac{3}{2}}(pq)^{\frac{1}{2}}} > \frac{c - npq}{(npq)^{\frac{1}{2}}}\right\}. \end{aligned} \tag{16}$$

Now as $c \to \infty$ let n vary as a function of c in such a way that $(c - npq)/(npq)^{\frac{1}{2}} \to -t$, and hence

$$n - c/(pq) \sim tc^{\frac{1}{2}}/(pq). \tag{17}$$

By the central limit theorem, $(q-p)(s_n - np)/(npq)^{\frac{1}{2}} \to N(0, (q-p)^2)$ in distribution, and $(s_n - np)^2/n^{\frac{3}{2}} \to 0$ in probability as $n \to \infty$. Hence letting Φ denote the standard normal distribution function, we have by (16) and (17)

$$\lim_{c\to\infty} P\left\{\frac{pq(N - c/(pq))}{c^{\frac{1}{2}}} \leqq t\right\} = \Phi(t/(q-p)).$$

A similar argument applies when $p < q$ to give

$$\lim_{c\to\infty} P\left\{\frac{pq(N - c/pq)}{|p-q|\,c^{\frac{1}{2}}} \leqq t\right\} = \Phi(t) \qquad (p \neq \tfrac{1}{2}). \tag{18}$$

In this case N is asymptotically $N(c/(pq), c(p-q)^2/(p^2q^2))$ in the usual sense as $c \to \infty$.

From (18) it is natural to conjecture that $\operatorname{Var} N \sim c(p-q)^2/(p^2q^2)$ as $c \to \infty$ $(p \neq \frac{1}{2})$. Straightforward modification of the arguments of Siegmund (1969) using (10) and Wald's identity for the fourth moment (Chow, Robbins & Teicher, 1965) show that in fact for all $p \neq \frac{1}{2}$ the difference $\operatorname{Var} N - c(p-q)^2/(p^2q^2)$ remains bounded as $c \to \infty$. No doubt it is also true that for u defined by (3), $Eu \to 0$, $\operatorname{Var} u \to 1$ as $c \to \infty$, although we have not verified the latter conjecture.

The asymptotic behaviour of the random sample size

$$M = \text{first } n \geqq 1 \text{ such that } s_n(n - s_n) > c$$

can be treated by the above methods with analogous results; M arises naturally in certain problems which involve minimizing the cost of observations cn plus a loss $(s_n/n - p)^2/(p^2q^2)$ due to error in estimating p.

References

Chow, Y. S., Robbins, H. & Teicher, H. (1965). Moments of randomly stopped sums. *Ann. Math. Statist.* **36,** 789–99.

Chung, K. L. (1968). *A Course in Probability Theory*. New York: Harcourt Brace and World.

Haldane, J. B. S. (1956). The estimation and significance of the logarithm of a ratio of frequencies. *Ann. Human Genetics.* **20,** 309–11.

Loève, M. (1963). *Probability Theory*. 3rd Ed. Princeton: Van Nostrand.

Siegmund, D. (1969). The variance of one-sided stopping rules. *Ann. Math. Statist.* **40,** 1074–7.

Modified likelihood estimation of a linear relationship

P. SPRENT
University of Dundee

SUMMARY

Structural and functional relationships are usually treated as distinct concepts, although point estimators of their parameters are often the same. Conventional maximum likelihood estimation leads to some anomalies. It is suggested that these might be avoided by considering certain functions of the observations and making inferences from these.

The proposed methods and their relation to other work are discussed. Although there is an element of arbitrariness in the choice of functions used for making inferences the consequences of a particular choice may be explored and the choice does not carry with it the subjective element involved in assigning a prior distribution for a Bayesian solution. Theoretical difficulties are mentioned but at this stage they are not explored in any detail.

1. INTRODUCTION

A characteristic of Pitman's work is that he has always sought sound theoretical bases for practical procedures. This is evident in his first major paper (Pitman, 1936) and in all his important writings since that date. He combines an ability to see the difficulties associated with a practical problem with the skill needed to produce the theory essential to its solution. In doing this he is always fully conscious of the value of what might be described as experimenting with ideas. This includes the study of *ad hoc* solutions to a problem, comparison with related problems and in general using past experience to cope with new situations. This paper is an experiment in ideas applied to some problems where existing solutions are not altogether satisfactory. Although it does not progress to the stage of providing elegant theories it is hoped that the ideas presented may stimulate interest in the theoretical background.

We consider first a set of n paired o' rvations (x_i, y_i) such that

$$\begin{aligned} x_i &= \xi_i + \delta_i, \\ y_i &= \eta_i + \varepsilon_i = \alpha + \beta\xi_i + \varepsilon_i, \end{aligned} \tag{1}$$

where the ξ_i, η_i are unknown parameters satisfying an unknown relationship $\eta_i = \alpha + \beta\xi_i$ and the δ_i, ε_i are independently distributed $N(0, \sigma_{00})$, $N(0, \sigma_{11})$ respectively for all i. The ξ_i have been termed incidental parameters by Neyman & Scott (1948) and the remaining parameters α, β, σ_{00}, σ_{11} are called the structural parameters. It is well known that the maximum likelihood solution to this problem runs into difficulties unless further assumptions are made. The problem is discussed by Lindley (1947), Madansky (1959), Kendall & Stuart (1967, ch. 29), Sprent (1969, ch. 3) and Moran (1971). Solari (1969) pointed out that the difficulty exists because the previously supposed maximum of the likelihood function is in reality a saddle point. A heuristic argument given by Kendall & Stuart demonstrates that the problem is incapable of satisfactory point estimate solution without further information or assumptions about error variances.

To supporters of the likelihood principle Solari's result is comforting. What had previously been described as a failure of the method has been shown in reality to be due to a deficiency in the problem and the likelihood function correctly reflects this.

However, even when additional assumptions are made so that the problem has a clear solution our difficulties with conventional maximum likelihood are not all over, for there may still be problems of consistency.

A commonly used additional assumption is to suppose the ratio $\lambda = \sigma_{11}/\sigma_{00}$ is known. The logarithm of the likelihood function corresponding to (1) with this additional assumption is

$$L^* = -\frac{n}{2}\ln\sigma_{00} - \frac{1}{2}\frac{\Sigma\,(x_i - \xi_i)^2}{\sigma_{00}} - \frac{1}{2}\frac{\Sigma\,(y_i - \alpha - \beta\xi_i)^2}{\lambda\sigma_{00}} + \text{const.} \tag{2}$$

It is easily shown that, on equating to zero the partial derivatives of L^* with respect to α, β, σ_{00} and the ξ_i, we get consistent estimators of α, β and the ξ_i, providing that in defining consistency we make an assumption of the type made by Lindley and El Sayyad (1968), namely that $\Sigma\,(\xi - \bar{\xi})^2/n$ converges to a finite limit, p, say, as n tends to infinity. Even with this assumption the estimator $\hat{\sigma}_{00}$ of σ_{00} converges to $\sigma_{00}/2$.

The inconsistency of $\hat{\sigma}_{00}$ is easily traced to the fact that the number of parameters $n+3$, is a function of the number of observations. Each paired observation introduces an additional ξ_i. In this situation there is no theory to guarantee consistency of maximum likelihood estimators analogous to that given by Wald (1949) for a fixed number of parameters.

In the next section we discuss some previous investigations that have enabled such inconsistencies to be avoided. Some of these ideas are applied to the present problem in §3 and to a more sophisticated example in §4. First we introduce the linear structural model and comment upon similarities to the functional model.

In the structural relationship model the observations have the form

$$\begin{aligned} x_i &= X_i + \delta_i, \\ y_i &= Y_i + \varepsilon_i = \alpha + \beta X_i + \varepsilon_i, \end{aligned} \tag{3}$$

where δ_i and ε_i have the properties described for (1) and are independent of the X_i, which are assumed $N(\mu, \sigma_{XX})$ where μ, σ_{XX} are unknown. Again, the problem is insoluble without further assumptions about the error variances. If we assume $\lambda = \sigma_{11}/\sigma_{00}$ is known the estimators of α and β are identical with those for the analogous functional relationship problem, but that for σ_{00} is twice its functional relationship analogue and is thus consistent. Standard results on consistency of maximum likelihood estimators now hold because the number of parameters is fixed.

It has long seemed intuitively reasonable to the present author that point estimators of the structural parameters α, β and σ_{00} might be the same for either model. Often in practice observations might well have arisen from either model and when the error structures are the same the particular method of generating the true underlying values seems irrelevant to the problem of point estimation. It is clear that the functional relationship model opens up the possibility of designed experiments as opposed to the taking of random observations and this gives scope for more precise estimates. The structural model must, for example, be imprecise if σ_{XX} is small compared to the error variances, but if x values corresponding to some appropriate set of ξ values may be chosen, then $\Sigma\,(\xi_i - \bar{\xi})^2/n$, the functional relation analogue of σ_{XX}, may be larger than the σ_{XX} associated with random observations and for a given error structure an increase in precision may be expected.

2. Inconsistency and choice of the likelihood function

An example of inconsistency of a maximum likelihood estimator often quoted in the literature is that in which we have two independent observations on each of n independent variates $X_1, X_2, \cdots, X_n$ where X_i is $N(\mu_i, \sigma_{00})$ and an estimator is required for the common variance σ_{00}. It is well known that the maximum likelihood estimators are

$$\hat{\mu}_i = x_{i.} = (x_{i1} + x_{i2})/2$$

and

$$\hat{\sigma}_{00} = \frac{1}{2n} \sum_i \sum_j (x_{ij} - x_{i.})^2.$$

Clearly the mean of the two observations on X_i is an unbiased estimator of μ_i, but the estimator $\hat{\sigma}_{00}$ converges to $\sigma_{00}/2$. Had the μ_i been known we would have obtained the consistent estimator

$$\tilde{\sigma}_{00} = \frac{1}{2n} \sum_i \sum_j (x_{ij} - \mu_i)^2,$$

and it is clear that our difficulties arise because we replace μ_i by its estimator $x_{i.}$.

In the discussion of Kalbfleisch and Sprott (1970), Barnard points out that he proposed a solution to this difficulty in 1946. Essentially it is to transform the observations to new variates

$$\begin{aligned} u_i &= x_{i1} - x_{i2}, \\ v_i &= x_{i1} + x_{i2}. \end{aligned} \tag{4}$$

The u_i are $N(0, 2\sigma_{00})$ and the v_i are $N(2\mu_i, 2\sigma_{00})$. In the absence of knowledge of the μ_i, the v_i contain no information on σ_{00}. However, the u_i give

$$\widehat{2\sigma_{00}} = \sum u_i^2/n$$

as an unbiased and consistent estimator of $2\sigma_{00}$. It is interesting to note that Barnard points out the possibility of obtaining information about σ_{00} from the v_i if we are prepared to make assumptions about the distribution of the μ_i and then use an empirical Bayes approach.

The inconsistency in the above example is a more extreme manifestation, resulting from the introduction of n parameters μ_i, of the bias exhibited in an even simpler situation discussed by Edwards (1972, p. 115).

He considers estimation of σ_{00} from a sample of n independent observations x_i from a $N(\mu, \sigma_{00})$ distribution. If μ is unknown the maximum likelihood estimator of σ_{00} is

$$\hat{\sigma}_{00} = \frac{1}{n} \sum (x_i - x_.)^2$$

which is consistent but biased. Had we known μ the maximum likelihood estimator

$$\tilde{\sigma}_{00} = \frac{1}{n} \sum (x_i - \mu)^2$$

would be consistent and unbiased. Clearly

$$\tilde{\sigma}_{00} = \hat{\sigma}_{00} + (x_. - \mu)^2$$

and the bias in $\hat{\sigma}_{00}$ is due to ignorance about the value of $(x_. - \mu)^2$ and represents the price we pay for replacing μ by $x_.$. In the absence of knowledge of μ this latter factor gives us no information about σ_{00}. Conventionally we overcome this difficulty by replacing the factor $1/n$ by $1/(n-1)$ and justify this by an argument based on degrees of freedom.

Edwards suggests an alternative approach. He points out that there exists a function of the observations, viz.

$$\frac{\Sigma\ (x_i - x_.)^2}{\sigma_{00}}$$

which is independent of μ and has a known distribution. Denoting $\Sigma\ (x_i - x_.)^2/n$ by s^2 we may write down the likelihood function for an observed s^2 in terms of the density function for the chi-squared distribution. The resulting maximum likelihood estimator is

$$\sigma^*_{00} = \frac{ns^2}{n-1}.$$

In this case the form of the sufficient statistic for σ_{00} has suggested an appropriate likelihood function for estimation of σ_{00}. As Edwards (p. 116) puts it—we restructure the problem so as to eliminate μ.

Plante (1971) adopts a similar approach and suggests elimination of nuisance parameters as just one method of modifying the likelihood principle in an objective way so that estimators with desirable properties may be obtained. His paper is mainly concerned with a finite dimensional parameter space. D. R. Cox recently pointed out to me that the idea of using a function of the observations rather than the observations themselves as the basis for a likelihood function appears to be due to Bartlett (1936).

Andersen (1970) discusses asymptotic properties of estimators obtained in the presence of many nuisance parameters by effectively eliminating them using conditional likelihoods. His results are applicable to the examples in this section but would not appear to be generally applicable to other sections of this paper since there do not appear to be minimal sufficient statistics for the incidental parameters of a functional relationship that are independent of the slope parameter.

The most closely related work is probably that by Kalbfleisch & Sprott (1970). They attempt to put modifications to the likelihood principle on a firm mathematical basis. Perhaps the most controversial ideas they introduce are their so-called Euclidean assumption and the concept of availability (or non-availability) of information. Further, questions of uniqueness arise in connection with their concepts of marginal and conditional likelihoods. This paper does not attempt to resolve these difficulties.

The author holds the view that there is no best principle of estimation applicable to all situations. What does seem important is to be aware of what assumptions one is making, and of their likely consequences. The aim of this paper is to demonstrate the workability of certain estimation procedures in functional relationship problems that are based upon the elimination or constraining of unwanted parameters.

3. Consistent estimators for a functional relationship

We return to the model given in (1) with the additional assumption that $\lambda = \sigma_{11}/\sigma_{00}$ is known. The log likelihood is given by (2). Equating partial derivatives to zero gives:

$$\Sigma\ (y_i - \hat{\alpha} - \hat{\beta}\hat{\xi}_i) = 0, \tag{5}$$

$$\Sigma\ \hat{\xi}_i(y_i - \hat{\alpha} - \hat{\beta}\hat{\xi}_i) = 0, \tag{6}$$

$$\Sigma\ (x_i - \hat{\xi}_i)^2 + \frac{\Sigma\ (y_i - \hat{\alpha} - \hat{\beta}\hat{\xi}_i)^2}{\lambda} = n\hat{\sigma}_{00}, \tag{7}$$

$$x_i - \hat{\xi}_i + \hat{\beta}\,\frac{y_i - \hat{\alpha} - \hat{\beta}\hat{\xi}_i}{\lambda} = 0, \tag{8}$$

$$i = 1, 2, \cdots, n.$$

We note that it is possible to obtain estimators of α, β and the ξ_i without the use of (7) and thus independently of the estimator of σ_{00}. It has already been indicated that substitution of the other estimators in (7) results in an inconsistent estimator of σ_{00}.

Had we known the ξ_i rather than having to estimate them, we would have had an equation similar in form to (7) with $\hat{\xi}_i$ replaced by ξ_i, viz.

$$\Sigma\ (x_i - \xi_i)^2 + \frac{\Sigma\ (y_i - \hat{\alpha} - \hat{\beta}\xi_i)^2}{\lambda} = n\hat{\sigma}_{00}. \tag{9}$$

This leads to a consistent estimator of σ_{00}. If we rewrite (9) in the form

$$\Sigma\ (x_i - \hat{\xi}_i + \hat{\xi}_i - \xi_i)^2 + \frac{\Sigma\ (y_i - \hat{\alpha} - \hat{\beta}\hat{\xi}_i + \hat{\beta}\hat{\xi}_i - \hat{\beta}\xi_i)^2}{\lambda} = n\hat{\sigma}_{00}$$

it is clear that the left hand side consists of terms on the left hand side of (7) together with terms containing either the factor $\hat{\xi}_i - \xi_i$ or $(\hat{\xi}_i - \xi_i)^2$. If the first factor only is involved the corresponding terms are easily shown to have expectation zero and terms involving the second factor are positive. Their contribution provides an explanation for the inconsistency of the estimator resulting from (7).

The n paired observations provide $2n$ data and we have to estimate $n + 3$ parameters. Of these, n are the incidental parameters ξ_i. One way to eliminate the latter is to replace our $2n$ data by n null variates of the form

$$u_i = y_i - \alpha^* - \beta^* x_i \tag{10}$$

where α^*, β^* are so chosen that the u_i contain only residual variation. The concept of a null variate is due to Williams (1955) and it is in a sense a generalization of the type of transformation u_i given in (4). Transformations of this type are also used in other situations, for example, in forming a set of orthogonal variates. An early example of the use of such transformations to ensure orthogonality is given by Pitman (1939).

The null variate in (10) differs from that in (4) in the sense that the u_i in (4) are completely known while in (10) α^* and β^* are unknown. However, since they are to be chosen so that the u_i contain only residual variation it follows that $E(u_i) = 0$ and $\mathrm{Var}(u_i) = (\lambda + \beta^{*2})\sigma_{00}$. If we write down a likelihood function based upon the u_i and differentiate with respect to α^*, β^* and σ_{00} to obtain maximum likelihood estimators of these parameters a somewhat unsatisfactory position arises owing to the interlocking of β^* and σ_{00} in the variance of the u_i. Indeed the estimator of β turns out to be the ordinary regression estimator for the regression of y on x which is known not to be consistent in this situation. Let us, however, consider the variate

$$w_i \frac{u_i}{\sqrt{(\lambda + \beta^{*2})}},$$

which, in view of the condition of nullity of the u_i is $N(0, \sigma_{00})$. The likelihood for a single observation of w_i has the form

$$L_1 = k\sigma_{00}^{-\frac{1}{2}} \exp\left\{ -\frac{1}{2}\frac{(y_i - \alpha^* - \beta^* x_i)^2}{(\lambda + \beta^{*2})\sigma_{00}} \right\}.$$

Writing down the likelihood for all the w_i and maximizing we get equations for the estimation of α^* and β^* that are independent of that for σ_{00} and are the same as those obtained by solving (5), (6) and (8) or by applying the method of generalized least squares (see, e.g., Sprent, 1966). This indicates that the precise values of the ξ_i are not relevant to the estimation of β. The estimator for σ_{00} is

$$\hat{\sigma}_{00} = \frac{1}{n} \frac{\Sigma\, (y_i - \hat{\alpha}^* - \hat{\beta}^* x_i)^2}{\lambda + \hat{\beta}^{*2}} \tag{11}$$

which is consistent. Note that if we assume β^* fixed the estimator (11) could also be obtained by equating the second sample moment of the u_i, viz. $\Sigma\; u_i^2/n$ to its expected value $(\lambda + \beta^{*2})\sigma_{00}$.

In this section we have followed principles implicit in the work of Andersen (1967) and that of Edwards and Plante already referred to in that we have where possible got estimators of parameters that are not dependent upon our estimators of other parameters. This was possible for β, but our estimator of σ_{00} still depends on β^*, though not upon the ξ_i. It is consistent but biased. This suggests following Edwards' approach to the simpler problem discussed in section 2 and basing our estimator on the relevant chi-squared statistic leading to

$$\hat{\sigma}_{00} = \frac{1}{n-2} \frac{\Sigma\, (y_i - \alpha^* - \beta^* x_i)^2}{\lambda + \beta^{*2}}$$

as an unbiased estimator.

Since each of the observed (x_i, y_i) introduces a new ξ_i it seems reasonable to suppose that there is only one function of x_i, y_i that will be free of the ξ_i and yet contain full information on the other parameters. The u_i seem an intuitively reasonable choice for this function but it is clear that some modification is still needed and that w_i is in fact the more appropriate choice. The likelihood function to which we are eventually led is essentially that obtained by Kalbfleisch & Sprott using their marginal likelihood concept. The reader should consult their paper for details. Andersen (1970) suggests that for even simpler models any transformation may result in loss of information or loss of efficiency. In the light of Kalbfleisch & Sprott's work it seems that without some such process there may indeed be a spuriously high apparent precision. However Andersen's results emphasize the need for care in choosing a transformation.

An interesting feature of the above approach is that it leads to estimators of α, β and σ_{00} exactly the same as those in the analogous structural relationship problem. The latter makes strong assumptions about the distribution of the true values. The functional relationship mode makes no distributional assumptions. We have shown that their actual distribution may well be irrelevant to point estimation. Bayesian attacks on this problem have often seemed unimpressive in their choice of prior distribution for the ξ_i. The above results suggest that this may be irrelevant. Empirical Bayes methods lead to verifiable approaches, but we do not consider these methods in this paper.

4. A SIMULTANEOUS FUNCTIONAL RELATIONSHIP PROBLEM

In the previous section we indicated the desirability of isolating the parameters to be estimated as far as this was possible. We achieved this either by elimination of unwanted parameters or by careful choice of a transformation to give estimating equations that did not involve unwanted parameters. There is no reason to believe that complete separation will always be possible and there then enters some arbitrariness in the choice of transformations.

We illustrate the situation for a problem considered recently by Barnett (1969), Williams (1969) and Moran (1971) as a structural relationship problem. The problem is of importance in calibration. Barnett uses maximum likelihood, but both he and Williams discuss an instrumental variate approach.

In the structural relationship model x_i, y_i, z_i are observed for n units where

$$x_i = X_i + \delta_i ,$$

$$y_i = Y_i + \varepsilon_i = \alpha_1 + \beta_1 X_i + \varepsilon_i ,$$

$$z_i = Z_i + \phi_i = \alpha_2 + \beta_2 X_i + \phi_i ,$$

where δ_i, ε_i, ϕ_i are independently distributed $N(0, \sigma_{00}), N(0, \sigma_{11}), N(0, \sigma_{22})$ respectively. The X_i have an unknown $N(\mu, \sigma_{XX})$ distribution and the problem is to estimate α_1, α_2, β_1, β_2 in the unknown relationships

$$Y = \alpha_1 + \beta_1 X ,$$

$$Z = \alpha_2 + \beta_2 X$$

between the random variables X, Y, Z. Interest may also attach to the estimation of μ, σ_{XX}, σ_{00}, σ_{11}, σ_{22}.

Unlike the case of a single bivariate relationship where we require further information about error variances or else a relaxation of normality assumptions to obtain a solution, there is now no difficulty. Since x_i, y_i, z_i have a trivariate normal distribution the nine parameters may be estimated using the three first and six second moments to provide the nine estimating equations. Writing

$$s_{00} = \Sigma (x_i - x_.)^2/n , \quad s_{01} = \Sigma (x_i - x_.)(y_i - y_.)/n ,$$

$$s_{12} = \Sigma (y_i - y_.)(z_i - z_.)/n , \text{ etc.,}$$

the maximum likelihood estimators were found by Barnett to be

$$\hat{\mu} = x_. , \quad \hat{\alpha}_1 = y_. - \hat{\beta}_1 x_. , \quad \hat{\alpha}_2 = z_. - \hat{\beta}_2 x_. , \quad \hat{\beta}_1 = s_{12}/s_{02} , \quad \hat{\beta}_2 = s_{12}/s_{01} , \quad \hat{\sigma}_{XX} = s_{01}s_{02}/s_{12} ,$$

$$\hat{\sigma}_{00} = s_{00} - s_{01}s_{02}/s_{12} , \quad \hat{\sigma}_{11} = s_{11} - s_{12}s_{01}/s_{02} , \quad \hat{\sigma}_{22} = s_{22} - s_{12}s_{02}/s_{01} ,$$

and these are all consistent.

We now consider the analogous functional relationship problem where

$$x_i = \xi_i + \delta_i,$$
$$y_i = \alpha_1 + \beta_1\xi_1 + \varepsilon_i,$$
$$z_i = \alpha_2 + \beta_2\xi_i + \phi_i,$$

where the X_i have been replaced by a mathematical variable ξ_i and we wish to estimate the parameters in relationships of the form

$$\eta = \alpha_1 + \beta_1\xi,$$
$$\zeta = \alpha_2 + \beta_2\xi.$$

We now have a problem solvable in terms of $n+7$ parameters and in attempting a maximum likelihood solution we obtain inconsistent estimators for $\sigma_{00}, \sigma_{11}, \sigma_{22}$. We do not give details of the algebra here but it is of some interest to consider a set of equations derivable after simple but tedious algebra from the maximum likelihood equations. If we carry over any relevant notation from the structural relationship model and in addition write

$$\hat{p} = \frac{1}{n}\Sigma\,(\hat{\xi}_i - \hat{\xi}_.)^2$$

the set of equations with unknowns $\hat{p}, \hat{\alpha}_1, \hat{\alpha}_2, \hat{\beta}_1, \hat{\beta}_2, \hat{\sigma}_{00}, \hat{\sigma}_{11}, \hat{\sigma}_{22}$ are

$$\hat{\alpha}_1 = y_. - \hat{\beta}_1 x_.\,, \tag{12}$$
$$\hat{\alpha}_2 = z_. - \hat{\beta}_2 x_.\,, \tag{13}$$
$$\hat{p} = s_{00} - \hat{\sigma}_{00}\,, \tag{14}$$
$$\hat{\beta}_1^2\hat{p} = s_{11} - \hat{\sigma}_{11}\,, \tag{15}$$
$$\hat{\beta}_2^2\hat{p} = s_{22} - \hat{\sigma}_{22}\,, \tag{16}$$
$$3 + \frac{\hat{\beta}_1 s_{01} - s_{11}}{\hat{\sigma}_{11}} + \frac{\hat{\beta}_2 s_{02} - s_{22}}{\hat{\sigma}_{22}} = 0, \tag{17}$$
$$\frac{s_{01} - \hat{\beta}_1 s_{00}}{\hat{\sigma}_{00}} + 3\hat{\beta}_1 + \frac{\hat{\beta}_2 s_{12} - \hat{\beta}_1 s_{22}}{\hat{\sigma}_{22}} = 0, \tag{18}$$
$$\frac{s_{02} - \hat{\beta}_2 s_{00}}{\hat{\sigma}_{00}} + \frac{\hat{\beta}_1 s_{12} - \hat{\beta}_2 s_{11}}{\hat{\sigma}_{11}} + 3\hat{\beta}_2 = 0. \tag{19}$$

These equations do not lead to Barnett's estimators of coefficients and error variances. However, if in (17), (18) and (19) we were to replace the terms 3, $3\hat{\beta}_1$ and $3\hat{\beta}_2$ by 2, $2\hat{\beta}_1$ and $2\hat{\beta}_2$ we would obtain Barnett's estimators together with $\hat{p} = s_{01}s_{02}/s_{12}$, giving a reasonable analogue of his $\hat{\sigma}_{XX}$. We now have a clue as to where the trouble lies. We have $3n$ data and $n+7$ parameters of which n are the ξ_i. This suggests that one third of our information is needed to estimate the ξ_i and that the diagonal term coefficients in (17), (18), (19) have been inflated as a consequence of our estimators $\hat{\sigma}_{ii}$ converging to $2\sigma_{ii}/3$.

Consider an approach using null variates. The obvious choice for these is

$$u_i = y_i - \alpha_1^* - \beta_1^* x_i,$$
$$v_i = z_i - \alpha_2^* - \beta_2^* x_i,$$

where $\alpha_1^*, \alpha_2^*, \beta_1^*, \beta_2^*$ are so chosen that u_i and v_i contain only residual variation. This implies

$$E(u_i) = E(v_i) = 0, \tag{20}$$

$$\text{Var}(u_i) = \sigma_{11} + \beta_1^{*2}\sigma_{00} = \sigma_{uu}, \text{ say}, \tag{21}$$

$$\text{Var}(v_i) = \sigma_{22} + \beta_2^{*2}\sigma_{00} = \sigma_{vv}, \text{ say}. \tag{22}$$

It follows from (21) and (22) that for fixed β_1^* and β_2^* maximum likelihood estimators of σ_{uu} and σ_{vv} are given by $\Sigma u_i^2/n$ and $\Sigma v_i^2/n$. Also, since u_i and v_i are correlated we have

$$\text{Cov}(u_i, v_i) = \beta_1^*\beta_2^*\sigma_{00} = \sigma_{uv}, \text{ say}, \tag{23}$$

where σ_{uv} may be estimated by $\Sigma\ u_i v_i/n$.

The problem of determining the appropriate 'fixed' values β_1^*, β_2^* remains unresolved, but several approaches suggest themselves. One possibility us to transform u_i and v_i still further to a pair of independent variates—a method used by Pitman (1939) in a very different problem. However, this does not help us to separate out the estimators of the coefficients and the error variances. Another approach is to consider the likelihood based on the joint distribution of the u_i and the v_i but here again the parameters are entangled in the estimating equations. Indeed, it seems impossible in this situation to disentangle the estimators of the slopes from those of the variances as was the case for β in section 3.

We note, however, that u_i rather than v_i gives information on β_1. Thus for β_1 it seems reasonable to consider a likelihood analogous to that based on w_i in section 3. However, it does not seem possible in this case to find a function whose distribution contains only one unknown parameter. For the appropriate choice of β_1^* to make u_i a null variate, the standardised variate

$$s_i = \frac{u_i}{\sigma_{11} + \beta_1^{*2}\sigma_{00}}$$

has a $N(0,1)$ distribution. The likelihood function based on the s_i thus has the form

$$L = k\exp\left\{-\frac{1}{2}\frac{\Sigma\ (y_i - \alpha_1^* - \beta_1^* x_i)^2}{(\sigma_{11} + \beta_1^{*2}\sigma_{00})}\right\}. \tag{24}$$

The value of the denominator in the exponent is constrained by the maximum likelihood estimator value of σ_{uu} and it follows from the constraints implied by (20) that $\hat{\alpha}_1 = y_. - \hat{\beta}_1 x_.$. Taking the logarithm of (24) and equating its derivative with respect to β_1^* to zero gives an equation involving $\hat{\beta}_1^*, \hat{\sigma}_{00}, \hat{\sigma}_{11}$. Defining a variate analogous to s_i based upon v_i and proceeding as above we get an equation in $\hat{\beta}_2^*, \hat{\sigma}_{00}, \hat{\sigma}_{22}$. These two equations may be combined with the ones formed by equating $\sigma_{uu}, \sigma_{vv}, \sigma_{uv}$ to their maximum likelihood estimators and all five equations solved for $\hat{\beta}_1^*, \hat{\beta}_2^*, \hat{\sigma}_{00}, \hat{\sigma}_{11}, \hat{\sigma}_{22}$. The solutions coincide with those obtained by Barnett in the structural analogue problem.

In (24) it may seem anomalous that we have written down a likelihood function based on observations having a known distribution, namely $N(0,1)$ in which there appear to be no parameters requiring estimation. The point is that there is an unknown parameter β_1^* contained within the observations. Effectively what we are doing is finding an estimator of this parameter subject to certain constraints upon the variances and covariances of the u_i and v_i and hence upon those of the x_i and y_i.

5. DISCUSSION

The ideas introduced in §§3 and 4 can be extended to other problems. For example, when there are replicated observations corresponding to each ξ_i for a single linear functional relationship the replication may be used to estimate σ_{00} and σ_{11}. The method of generalized least squares discussed by Sprent (1966) uses these estimates in place of the true values and this method gives consistent estimators of α and β. A full maximum likelihood treatment of the problem leads to inconsistencies that are easily traced to the presence of the unknown ξ_i. In this case, if we have no knowledge of the ξ_i we are led to estimators of the variances based upon replication as appropriate estimators.

The estimators introduced on intuitive and exploratory grounds in this paper turn out to be very similar to those obtained by Kalbfleisch & Sprott. It is clear in the problem considered in §4 that we might proceed using functions other than s_i. In fact, two other possibilities were briefly mentioned in that section but rejected. Once we supplement the likelihood principle in this way we lose the uniqueness of estimators. This is part of the price we pay for introducing consistency and other desirable properties. Andersen's (1970) results suggest that we may also pay a price in loss of information and efficiency. It is of course possible to compare estimators and thus to determine which have generally desirable properties. Clearly more work is needed on the theoretical aspects of the approach given here and the same is true with the more basic work of Edwards, Plante and Kalbfieisch Sprott discussed in §2.

In particular we require a useable definition of *available information* in the presence of nuisance parameters. We need to specify more clearly our priorities for desirable properties of estimators, e.g. consistency, efficiency, sufficiency, unbiasedness, etc..

The arbitrary element in the choice of a null variate suggests as a name for the class of processes described in this paper *empirical likelihood*. The process allows more flexibility than is possible with rigid adherence to the likelihood principle. It is possible to assess the consequences of any particular choice of a null variate.

It is to be hoped that just as empirical Bayes methods have enabled us to avoid some of the less desirable subjective judgements associated with pure Bayesian inference, so the approaches suggested in this paper may introduce some flexibility that is missing in a too rigid adherence to the likelihood principle.

ACKNOWLEDGEMENTS

Research leading to this paper was carried out in part while the author was a visiting research fellow at the University of Queensland in July 1972. His thanks are due to Professor S. Lipton for an invitation to visit Queensland and to the head of the Department of Mathematics at the University of Queensland, Professor C. S. Davis, for supporting the visit.

REFERENCES

ANDERSEN, E. B. (1967). On partial sufficiency and partial ancillarity. *Skand. Aktuarietidskr.* **50**, 137–52.

ANDERSEN, E. B. (1970). Asymptotic properties of conditional maximum likelihood estimators. *J. Roy. Statist. Soc.* B **32**, 283–301.

BARNETT, V. D. (1969). Simultaneous pairwise linear structural relationships. *Biometrics*, **25**, 129–42.

BARTLETT, M. S. (1936). The information available in small samples. *Proc. Camb. Phil. Soc.* **32**, 560–6.

EDWARDS, A. W. F. (1972). *Likelihood.* Cambridge University Press.

KALBFLEISCH, J. D. & SPROTT, D. A. (1970). Application of likelihood methods to models involving large numbers of parameters. *J. Roy. Statist. Soc.* B **32,** 175–208.

KENDALL, M. G. & STUART, A. (1967). *The Advanced Theory of Statistics*, vol. 2, 2nd edition. London: Griffin.

LINDLEY, D. V. (1947). Regression lines and linear functional relationships. *J. Roy Statist. Soc. Suppt.* **9,** 218–44.

LINDLEY, D. V. & EL SAYYAD, G. M. (1968). The Bayesian estimation of a linear functional relationship. *J. Roy. Statist. Soc.* B **30,** 190–202.

MADANSKY, A. (1959). The fitting of straight lines when both variables are subject to error. *J. Amer. Statist. Assoc.* **54,** 173–205.

MORAN, P. A. P. (1971). Estimating structural and functional relationships. *J. Multivariate Anal.* **1,** 232–55.

NEYMAN, J. & SCOTT, E. L. (1948). Consistent estimates based on partially consistent observations. *Econometrica* **16,** 1–32.

PITMAN, E. J. G. (1936). Sufficient statistics and intrinsic accuracy. *Proc. Camb. Phil. Soc.* **32,** 567–79.

PITMAN, E. J. G. (1939). A note on normal correlation. *Biometrika* **31,** 9–12.

PLANTE, A. (1971). Counter examples and likelihood. In *Foundations of Statistical Inference*, ed. by V. P. Godambe and D. A. Sprott, 357–71. Toronto: Holt, Rinehart and Winston.

SPRENT, P. (1966). A generalized least-squares approach to linear functional relationships. *J. Roy. Statist. Soc.* B **28,** 278–97.

SPRENT, P. (1969). *Models in Regression and Related Topics*. London: Methuen.

SOLARI, M. E. (1969). The 'maximum likelihood solution' of the problem of estimating a linear functional relationship. *J. Roy. Statist. Soc.* B **31,** 372–75.

WALD, A. (1949). A note on the consistency of the maximum likelihood estimate. *Ann. Math. Statist.* **20,** 595–601.

WILLIAMS, E. J. (1955). Significance tests for discriminant functions and linear functional relationships. *Biometrika* **42,** 360–81.

WILLIAMS, E. J. (1969). Regression methods in calibration problems. *Bull. Int. Statist. Inst.* **43,** 17–27.

Optimal invariant tests for uniformity

G. S. WATSON
Princeton University

SUMMARY

The Kolmogorov and Cramér-von Mises families of tests for uniformity on the unit interval are not derived as optimal tests. However on the circle and its generalizations, it is possible to derive optimal invariant tests for uniformity. Beran (1968) studied tests of local alternatives. These have the integral square form. Watson's (1961) circular variant of the Cramér-von Mises test was shown to be optimal in a certain situation. In this paper, tests for distant alternatives are derived. These have the supremum form. Circular variants of D_n^+ and D_n^- are shown to be optimal. Efforts to show that Kuiper's (1960) statistic and the obvious circular variant of Kolmogorov's test are optimal failed. Some new distribution questions are raised.

1. INTRODUCTION

The writer was privileged to hear the lectures of E. J. G. Pitman at the University of North Carolina in 1948/9 on Non-parametric Statistics, Inference and Applied Probability. In the latter course Pitman stressed the fundamental nature of the uniform distribution on $(0,1)$. In particular the Kolmogorov (K_n) and Cramér-von Mises (W_n^2) tests were discussed. In the former course the concept of the relative asymptotic efficiency of tests was first developed. A variant of this idea due to Bahadur (1967) has been used (e.g. Abrahamson (1967)) to compare tests like K_n and W_n^2. Such tests have never been derived from test theory, merely compared. Tests, and estimation theory, were the subject matter of Pitman's other course of lectures. One of his contributions to this subject was the observation that tests and estimators should behave simply under changes of location and scale in the problem—that is, invariance.

The aim of this paper is to point out that, on the circle (and its generalizations), these interests of Pitman fit together beautifully. The bulk of the work has already been done by Beran (1968) in his discussion of best invariant tests for "local" alternatives to uniformity. This will be briefly described at the beginning of §2 as a necessary prelude to the formulation of the same problem for distant alternatives, the best invariant tests for which will turn out to be likelihood ratio tests. In §3 some examples of local and distant tests are given.

In §4, there is a discussion, for the circle, of tests closely analogous to K_n and W_n^2. Since the cumulative distribution function is rather unnatural except on the real line, one should not expect it to appear in optimal tests on the circle, for example. Density functions are appropriate on any space and so they appear in the optimal test statistics. On the circle these tests only involve the sample distribution function for one very special type of alternative.

Portnoy & Stein (1971) have pointed out difficulties with invariant tests when R^k is the sample space. Epp (1971) has shown that these do not appear for the compact sample spaces we consider below.

2. Local and distant alternatives to uniformity

The idea of using optimal invariant tests on the unit circle is due to Ajne (1968). Let the perimeter be of length unity and the alternative to uniformity be a density equal to $2q$ on a semi-circle and $2(1-q)$ on the other semicircle, $1 > q > \frac{1}{2}$. Since the semi-circles are not fixed in position, the test should be invariant under change of position. He derived the optimal invariant statistics

$$A = \max_x N(x), \qquad (q \text{ near } 1), \tag{1}$$

$$B = \frac{1}{n}\int_0^1 \left(N(x) - \frac{n}{2}\right)^2 dx, \; (q \text{ near } \tfrac{1}{2}), \tag{2}$$

where $N(x)$ is the number of observations in the arc $[x, x + \frac{1}{2}]$.

(1) is an example of a *distant* alternative and (2) of a *local* alternative to uniformity. One may generalize by considering different alternatives and by leaving the circle for a sphere or further, for a compact homogeneous space. Beran carried this out for (2). This paper suggests the program for (1). Also it is clear that there is some correspondence between K_n and A and between W_n^2, U_n^2 and B, where U_n^2 is the invariant uniformity test statistic on the circle suggested by Watson (1961).

A probability density $h(x)$ on a circle is a nonnegative function of integral and period 1. On the null hypothesis $h(x) = 1$. If the alternative density is $f(x + \phi)$, for an unspecified ϕ, and the observations are $x_1, x_2, \cdots, x_n$, the best invariant test rejects the hypothesis of uniformity for large values of

$$\int_0^1 \prod_1^n f(x_i + \phi) d\phi . \tag{3}$$

This is proved in e.g. Lehmann (1959). If for all x, $|f(x) - 1| < \varepsilon$, the alternative is said to be *local*. Then the test (3) is effectively: reject for large values of

$$\frac{1}{n}\int_0^1 \left\{ \sum_1^n [f(x_i + \phi) - 1]\right\}^2 d\phi , \tag{4}$$

a generalization of (2). The divisor n is introduced so that (4) has a nontrivial asymptotic distribution. If instead of a circle we take a compact topological group G, observations $g_1, \cdots, g_n$, invariant measure dg, $\int_G dg = 1$, a density $h(g) = 1$ on the null hypothesis, and $h(g) = f(kg)$, $k \in G$ on the alternative, the same arguments yield the test statistic

$$\frac{1}{n}\int_G \left\{ \sum_1^n [f(gg_i) - 1]\right\}^2 dg . \tag{5}$$

For the sphere, the location group G (the rotation group $R(3)$) and the sample space X do not have the same structure, as they do on the circle. But each point x on the sphere can be associated with the set of rotations leaving it unchanged. Beran reduces the problem to the previous case and shows that the local statistic is the same as (5) but with x_i in place of g_i.

Thus the local tests come from examining (3) or its generalizations where f is near 1 all over the sample space and lead to "integral square" statistics. Their derivation above may be rephrased: look at the sequence of best invariant tests of uniformity versus a density $1 + \delta(f(x + \phi) - 1)$, when $\delta \to 0$ and f is a bounded density, defined by

$$\int_0^1 \prod_1^n \{1 + \delta[f(x_i + \phi) - 1]\} d\phi . \tag{6}$$

The alternative most *distant* from the uniform distribution is surely a distribution concentrated at one point. To get a best distant test, we should then consider a sequence of alternatives concentrating more and more around one point.

For the circle, suppose $f(x)$ is a unimodal density function on $(0, 1)$ and defined elsewhere periodically and consider the sequence of densities $\kappa_p^{-1} f(x)^p$ where $\kappa_p = \int_0^1 f(x)^p dx$ and $p \to \infty$. If $f(x)$ is reasonably well behaved, it is intuitively clear that this sequence tends to a distribution wholly centred on the mode of $f(x)$. But the sequence will always lead to distributions more and more distant from uniformity so f will not be restricted to lead to a limit distribution entirely at one point. Finally $(\kappa_p)^{1/p}$ tends to the essential supremum of f, that is, the supremum of f except for sets of measure zero.

Using this power sequence in (3), we get analogously with (6) a sequence of best tests defined by large values of

$$\int_0^1 \prod_1^n \frac{f(x_j + \phi)^p}{\kappa_p} d\phi ,$$

or equivalently

$$\left(\int_0^1 (\Pi f(x_j + \phi))^p d\phi \right)^{1/p} .$$

As $p \to \infty$, limit of this sequence is the test which rejects when

$$\operatorname*{ess\ sup}_{0 \leq \phi \leq 1} \prod_1^n f(x_j + \phi) \text{ is large}$$

or

$$\operatorname*{ess\ sup}_{0 \leq \phi \leq 1} \sum_1^n \log f(x_j + \phi) \text{ is large.}$$

Thus we have generated a class of best invariant tests of uniformity on the circle, appropriate for distant alternatives. (7) is of the same type as (1), i.e., like A and Kolmogorov's K_n it involves a supremum. To find the significance points so that they may be used in practice, we need the distribution of ess sup $\log f(x_j + \phi)$ when the x_j's are uniform. But (7) is the logarithm of the likelihood ratio (L.R.) test statistic for testing uniformity versus $f(x + \phi)$. However it does not seem possible to call on L.R. theory to get the properties of the test. More interesting is the question of why a discussion of distant alternatives has led to this L.R. test. We have no answer. Although in some cases it may be possible to find the distribution for finite n, asymptotic approximations will be the rule. As $n \to \infty$, the statistic, being approximately $n \int_0^1 \log f(x) dx$, tends to $-\infty$. Thus it is necessary to standardize. The examples of §3 show that a variety of distributions may result. They may all belong to one family as is the case of the local tests. There Beran was able to use and generalize a Fourier series method due to the writer to reduce all statistics to infinite quadratic forms. Finding the distribution in each case is then a matter of recognizing and inverting the characteristic function. However we have no general method for solving the distribution problem raised by (7). It is intuitively clear that it may be reduced to finding the distribution of the supremum of a stationary Gaussian process on the circle.

We conclude this section by stating the generalization of (7) which rejects when

$$\sup_{g \in G} \Sigma \log f(gx_i) \tag{8}$$

is large. Here $x_1, \cdots, x_2$ are the observations in the homogeneous space X which admits the transitive transformation group G.

3. EXAMPLES

Example 1. The von Mises density on the circle is $[2\kappa I_0(\kappa)]^{-1} \exp\{\kappa \cos 2\pi(x-\xi)\}$, $\kappa \geqq 0$. For $\kappa = 0$, it is the uniform density and as $\kappa \to \infty$, the distribution concentrates on ξ which is always the "mean" direction. It is well known (see e.g. the book by Mardia, 1972) that the globally best test of $\kappa = 0$ rejects when the length of the sample resultant is too long, i.e. $R^2 = (\Sigma \cos 2\pi x_i)^2 + (\Sigma \sin 2\pi x_i)^2$ is too large. If this density is taken for f in (7), (7) reduces, as it must, to a test based on

$$\sup \Sigma \cos 2\pi(x_i - \xi) = R. \tag{9}$$

The pth power of the von Mises is again von Mises with parameter κp so fits our intuitive requirement of leading to a concentrated distribution. The asymptotic distribution is provided by the old fact that

$$\frac{2R^2}{N} \sim \chi_2^2. \tag{10}$$

Finally in this example it is easy to verify that the local test (4) reduces to the large R test.

Example 2. For Ajne's distribution, we may write

$$f(x - \theta) = 2(1-q)(q/(1-q))^{I[\theta,\theta+\frac{1}{2}](x)}, \ (1 > q > \tfrac{1}{2}) \tag{11}$$

where $I_{[\theta,\theta+\frac{1}{2}]}(x)$ is the indicator function of the interval $[\theta, \theta + \frac{1}{2}]$. Thus substitution in (7) gives

$$n \log 2(1-q) + \log\left(\frac{q}{1-q}\right) \sup_{\theta} \sum_{j=1}^{n} I_{[\theta,\theta+\frac{1}{2}]}(x_j)$$

But $q > \frac{1}{2}$ and $\Sigma_j I_{[\theta,\theta+\frac{1}{2}]}(x_j) = N(\theta)$ as defined below equations (1), (2), so that test (1) is rederived. Ajne's distribution does not satisfy the intuitive requirements since the pth power tends to $2I_{[\theta,\theta+\frac{1}{2}]}(x)$ as $p \to \infty$. This is however the most distant member of the Ajne family from uniformity, corresponding to $q = 1$. Ajne (1968) has given both the finite and asymptotic distributions of A. The latter is

$$\lim_{n \to \infty} \operatorname{Prob}\left(\frac{2A - n}{\sqrt{n}} \geqq c\right) = 2c\left(\frac{2}{\pi}\right)^{\frac{1}{2}} \sum_{j=0}^{\infty} \exp(-\tfrac{1}{2}(2j+1)^2 c^2). \tag{12}$$

This is a very different kind of result from (10). Here (4) reduces to (2) so that, unlike Example 1, the local and distant tests do not coincide. There is no U.M.P. test of uniformity as in Example 1.

Example 3. For the Fisher distribution on the sphere, (8) leads immediately to the large R test since it is the three-dimensional version of Example 1. The asymptotic approximation is

$$\frac{3}{N}R^2 \sim \chi_3^2.$$

Example 4. Another example of (8) is provided by a circular multinomial or polygon with m vertices. The null hypothesis is probability $1/m$ at each vertex. The alternative is that subject to an unknown cyclic permutation the vertex probabilities are $p_0, p_1, \cdots, p_{m-1}$. If one p_i is greater than all the others, powering leads to all the probability at one vertex. If the observed frequencies at vertices $0, 1, \cdots, m-1$ are $N(i)$, $i = 0, \cdots, m-1$ and $p(i + k \bmod m)$ denotes the probabilities moved ahead k vertices, $k = 0, \cdots, m-1$, then (8) may be written

$$\max_k \sum_0^{m-1} N(i) \log p(i + k \bmod m). \tag{12}$$

For tests, we require the distribution of (12) when the $N(i)$ are the frequencies from n observations of an m-nomial with equal cell probabilities. The distribution is not known. If one probability is much greater than all the others which are equal, (12) reduces to $\max_i N(i)$. If the odd probability is the smallest, it reduces to $\min_i N(i)$. Tests closely related to these have been computed by Dudley (1971).

4. Analogues of K_n and W_n^2 on the circle

Kuiper (1960) suggested a test of uniformity based on large values of

$$V_n = \sup \sqrt{n}(F_n(x) - x) - \inf \sqrt{n}(F_n(x) - x)\,; \tag{13}$$

he showed V_n did not depend on the choice of the origin for forming the sample distribution function, and gave the asymptotic distribution of V_n. Watson (1961) suggested the invariant statistic

$$U_n^2 = n \int_0^1 \left\{F_n(x) - x - \int_0^1 (F_n(y) - y)dy\right\}^2 dx \tag{14}$$

and gave its asymptotic distribution. Beran showed that U_n^2 is a locally optimal invariant test of type (4) if f has Fourier coefficients c_m such that $|c_m| \propto m^{-1}$.

A density function on $(0, 1)$ with this property is

$$g(x) = (1 - cx)(1 - c/2)^{-1} \quad (c \leqq 1). \tag{15}$$

It may be shown that

$$g(x) = 1 + \frac{2c}{(2-c)\pi} \sum_1^\infty \frac{\sin 2\pi m x}{m} \tag{16}$$

and that

$$\sum_1^n g(x_j - \phi) - 1 = \frac{-2cn}{2-c}(F_n(\phi) - \phi - \tfrac{1}{2} + \bar{x}). \tag{17}$$

Furthermore these last expressions are unchanged if any number, say ξ, is added to all the x_j and ϕ. The proof and origin of these results lie in the method of discovery of U_n^2. This was to express $\sqrt{n}(F_n(\phi) - \phi)$ as a Fourier series $\Sigma_{-\infty}^{\infty} c_m e^{2\pi i m\phi}$. The term $c_0 = \int_0 \sqrt{n}(F_n(\phi) - \phi)d\phi = \sqrt{n}(\frac{1}{2} - \bar{x})$ so that

$$\sqrt{n}(F_n(\phi) - \phi) - \sqrt{n}(\tfrac{1}{2} - \bar{x}) = \sum_{-\infty}^{\infty}{}' \left(\frac{-1}{2\pi i m} \frac{1}{\sqrt{n}} \sum_j e^{2\pi i m x_j}\right) e^{-2\pi i m\phi}, \tag{18}$$

This makes it clear why Kuiper's V_n and U_n^2 are invariant. It also suggests that

$$\sup_{\phi} (F_n(\phi) - \phi - \tfrac{1}{2} + \bar{x}) \quad \text{or} \quad \inf_{\phi}(F_n(\phi) - \phi - \tfrac{1}{2} + \bar{x}) \tag{19}$$

might be used as invariant test statistics. The following argument shows that these tests may be derived from (7). They are the analogues of D_n^+ and D_n^- on the circle.

For the density $f(x + \phi)$ in (7), consider

$$f(x + \phi) = Ke^{g(x+\phi)}$$

where $g(x + \phi)$ is (16) continued periodically and K is the normalization constant, a function of c ; f is a unimodal density so the test (7) is an optimal invariant test of a distant hypothesis. But from (7) and (17) we are led to

$$\sup \left\{n - \frac{2cn}{2-c}[F_n(\phi) - \phi - \tfrac{1}{2} + \bar{x}]\right\}. \tag{20}$$

Depending on the sign of c we are led to either of the tests (19). It does not seem possible to derive V_n or the analogue of K_n

$$\sup | F_n(\phi) - \phi - \tfrac{1}{2} + \bar{x} |$$

this way, although one feels they are tests which try to cope with the possibility that c may have either sign.

Defining

$$M_n = \sup_{\phi} \sqrt{n}(F_n(\phi) - \phi - \tfrac{1}{2} + \bar{x}),$$

a Monte Carlo experiment for $n = 20$ showed that, to surprisingly good approximation,

$$M_n = \tfrac{1}{2}V_n + u$$

where u is independent of V_n, normal with zero mean and a variance which increases slowly with V_n to a value of .0095. Using the distribution of Kuiper's V_n and assuming u is small compared to V_n it is possible to find a formula for Prob$(M_n > z)$ and so to estimate significance points a little more accurately than from direct Monte Carlo. The details will be reported elsewhere.

The writer is grateful to R. J. W. Beran for comments and to David Hunter for some exploratory computing. The work was partially supported by the Office of Naval Research under Contract NONR4010(09) to the Department of Statistics, Princeton University.

REFERENCES

ABRAHAMSON, I. G. (1967). Exact Bahadur efficiencies for the Kolmogorov-Smirnov and Kuiper one-and two-sample statistics. *Ann. Math. Statist.* **38**, 1475–90.

AJNE, B. (1968). A simple test for uniformity of a circular distribution. *Biometrika* **55**, 343–54.

BAHADUR, R. R. (1967). Rates of convergence of estimates and statistics. *Ann. Math. Statist.* **38,** 303–25.

BERAN, R. J. W. (1968). Testing for uniformity on a compact homogeneous space. *J. Appl. Prob.* **5,** 177–95.

DUDLEY, R. M. (1971). Binomial tests for goodness of fit. Unpublished Report, Mathematics Department, M.I.T.

EPP, R. (1970). Invariant tests on compact spaces. Unpublished.

KUIPER, N. (1960). Tests concerning random points on a circle. *Proc. Kon. Ned. Akad. van Wettenschappen, A* **63,** 38–47.

LEHMANN, E. L. (1959). *Testing Statistical Hypotheses.* New York: Wiley.

MARDIA, K. V. (1972). *Statistics of Directional Data.* London and New York: Academic Press.

PORTNOY, S. & STEIN, C. (1971). Inadmissibility of the best invariant test in three or more dimensions. *Ann. Math. Statist.* **42,** 799–801.

WATSON, G. S. (1961). Goodness-of-fit tests on a circle. *Biometrika* **48,** 109–14.

The use of mathematical models in enzyme kinetics

By M. J. ATKINSON
University of Adelaide

1. Mathematical models and data analysis

Enzymes are naturally occurring organic compounds which increase the rates of certain chemical reactions while themselves remaining unchanged. Enzyme kinetics, the study of the mechanisms of these catalytic reactions, is concerned with the formulation and testing of possible models of a reaction and the estimation of the unknown parameters involved.

A postulated mechanism may be expressed in terms of stoichiometric equations (1) or equivalently as a set of differential equations (3) relating rates of change of concentration of the reactants to the actual concentrations. The velocity constants of the individual reactions in the mechanism and the initial concentrations of the reactants enter into the equations also, and it is the estimation of the unknowns among these which presents the problem.

The kinetic data which are to be used for fitting the model usually come in the form of automatically recorded graphs relating concentration of one of the reactants to time elapsed. The shape of the curve depends upon the underlying mechanism and also upon the initial concentration of reactants.

Because the model equations do not have explicit solutions, biochemists have traditionally made simplifying assumptions (e.g. that concentration of intermediates is constant) which permit the derivation of more tractable equations by eliminating time as a variable. For example Michaelis & Menten (1913) derived an equation relating the initial slope of the graph to the initial concentration of the substrate for a simple one substrate—one product reaction.

By such means the model may be tested to a certain extent and simple functions of the velocity constants may be estimated. Data must be in the form of initial slopes of the recorded curves, and inaccuracy in measuring these is often a fairly serious source of error. Also, only one value is used from each experiment and the rest of the information in the curve is discarded.

The availability of high speed computing facilities permits fresh approaches to the problem of testing and estimation, using the data curves in their original form.

One possibility is that, making the same simplifying assumptions as before, an explicit solution to the model equations may be found (7), which can be fitted to the data by some non-linear least squares method. The drawbacks of this technique are that there are fewer estimated parameters than the number of unknown velocity constants, that the parameters are not simple functions of these velocity constants, and that, since the recorded graph is the result of a single experiment, points read from the curve are not independent. Hence the application of standard statistical techniques is of doubtful validity.

The most direct method of getting values of the velocity constants is to discard the ideal of statistical estimation and solve the differential equations numerically. Using the best values that the biochemist can provide by other means as the starting values for the velocity constants, adjustments can be made to these until the numerical solution fits the data curve satisfactorily. This requires a subjective judgement as to the size and nature of the deviations which can be tolerated. Guidance in this can be obtained by performing duplicate experiments to give an indication of the type of variations which may occur in the data curve due to experimental error.

When a satisfactory set of parameter values has been arrived at, from the point of view of correspondence between the data and the numerical solution, it is necessary to test whether the fitted model is in fact applicable. The first requirement is that the estimated values of the unknown concentrations and velocity constants must be physically possible. If that is so, the model may be tested further by using it to predict the course of reactions with different initial concentrations of reactants. The fitting of a single simple curve by a model system involving more than three parameters is hardly convinving evidence in support of the model, but if a family of curves can be fitted using the same parameter values throughout, its credibility becomes much greater.

Such numerical solutions do not of course yield estimates of the variances and covariances of the parameter estimates. However, the process of adjustment of estimates and recalculation of the solution will provide information about the sensitivity of the model to changes in the parameter values and this, together with a knowledge of the experimental variation, will permit the construction of confidence intervals based on subjective probabilities. These intervals will have the property that, if the value of the parameter lies within the interval, then the calculated solution will be indistinguishable from the data within the limits of experimental error.

In certain favourable cases a quadratic approximation may be fitted to a suitable measure of the deviations of the calculated solutions from the data. By this means an approximation to the variance-covariance matrix of the parameter estimates may be obtained.

2. EXAMPLE; MECHANISM OF ACTION OF LIVER ALCOHOL DEHYDROGENASE

Liver alcohol dehydrogenase catalyses the reversible transfer of hydrogen from ethanol (CH_3CH_2OH) to nicotinamide-adenine dinucleotide (NAD), producing reduced NAD (NADH) and acetaldehyde (CH_3CHO).

Experimental method: The action of liver alcohol dehydrogenase was investigated under the following conditions: 23°C, 0.05 μM phosphate buffer (pH 7.15), 0.1 M ethanol.

The enzyme used was Calbiochem A grade horse liver alcohol dehydrogenase, 10 mg/ml crystalline in 0.02 M phosphate buffer. Ten microlitres of this solution were added to the reaction mixture in a 3 ml cuvette to provide an enzyme concentration of 33 μ g/ml in the final volume.

The course of the reaction was followed using a UNICAM SP 800 recording spectrophotometer. Because of the small scale of the chart used with this instrument the result was also recorded on a Servoscribe chart recorder, which permitted expansion of the scales of both optical density and time.

Duplicate families of data curves were obtained by varying the initial concentration of NAD (40, 60, 100, 120, 160, 240 μM), while holding all other factors constant as above.

Formulation of the model. Theorell and his coworkers (Theorell & Bonnichsen, 1951;

Theorell Chance, 1951; Theorell, Nygaard & Bonnichsen, 1955) have formulated the following reaction mechanism for this enzyme:

$$\begin{aligned} \mathrm{NADH + E} &\underset{k_2}{\overset{k_1}{\rightleftharpoons}} \mathrm{NADH.E} \\ \mathrm{NADH.E + CH_3CHO + H^+} &\underset{k_6}{\overset{k_4}{\rightleftharpoons}} \mathrm{NAD.E + CH_3CH_2OH} \\ \mathrm{NAD.E} &\underset{k_5}{\overset{k_3}{\rightleftharpoons}} \mathrm{NAD + E}. \end{aligned} \qquad (1)$$

Because of the nature of chemical change, the following constraints are implied:

$$\begin{aligned} [\mathrm{NAD}] + [\mathrm{NAD.E}] + [\mathrm{NADH.E}] + [\mathrm{NADH}] &= C_1 \\ [\mathrm{E}] + [\mathrm{NAD.E}] + [\mathrm{NADH.E}] &= C_2 \\ [\mathrm{CH_3CH_2OH}] + [\mathrm{CH_3CHO}] &= C_3 \\ [\mathrm{CH_3CHO}] &= [\mathrm{NADH}] + [\mathrm{NADH.E}]. \end{aligned} \qquad (2)$$

Square brackets indicate concentrations. C_1, C_2 and C_3 are constants.

Putting

x = concentration of NADH

w = concentration of NADH.E

u = concentration of CH_3CHO

z = concentration of NAD.E

v = concentration of CH_3CH_2OH

y = concentration of NAD

e = concentration of enzyme,

application of the Law of Mass Action to equations (1) gives the set of differential equations:

$$\begin{aligned} \frac{dx}{dt} &= -k_1ex + k_2w \\ \frac{dw}{dt} &= k_1ex - k_2w - k_4wu + k_6zv \\ \frac{du}{dt} &= -k_4wu + k_6zv \\ \frac{dz}{dt} &= k_4wu - k_6zv - k_3z + k_5ey \\ \frac{dv}{dt} &= k_4wu - k_6zv \\ \frac{dy}{dt} &= k_3z - k_5ey. \end{aligned} \qquad (3)$$

Assuming that the course of the reaction is to be studied in a system where initially only NAD, ethanol and enzyme are present, the equations (2) may be written as:

$$\begin{aligned} x + w + z + y &= y_0 \\ e + w + z &= e_0 \\ u + v &= v_0 \\ u &= x + v, \end{aligned} \tag{4}$$

where y_0, e_0 and v_0 are the initial concentrations of NAD, enzyme and ethanol respectively.

Since the data obtained for this reaction consisted of the sum of NADH and NADH.E, the variable, $\theta = x + w$ was introduced, and x was eliminated to give, after simplification, the following reduced system of equations

$$\begin{aligned} \frac{d\theta}{dt} &= -k_4 w\theta + k_6(y_0 - \theta - y)(v_0 - \theta) \\ \frac{dw}{dt} &= k_1(\theta - w)(e_0 - (y_0 - \theta - y) - w) - k_2 w + \frac{d\theta}{dt} \\ \frac{dy}{dt} &= k_3(y_0 - \theta - y) - k_5 y(e_0 - (y_0 - \theta - y) - w). \end{aligned} \tag{5}$$

Steady state solution. If one assumes that the steady state has been reached, then both dw/dt and dz/dt may be set equal to zero, and the equations (5) can be simplified since now w and z are constants with values w_s and z_s respectively. Hence, from equations (5) we have,

$$d\theta/dt = -k_4 w_s \theta + k_6 z_s(v_0 - \theta). \tag{6}$$

The solution of this gives an equation of the form

$$\theta = a(1 - e^{-b(t+k)}). \tag{7}$$

This approximate solution was fitted to the data by non-linear least squares, employing the technique of Nelder & Mead (1965). Although a good fit was obtained over most of the range, the deviation of the fitted function from the data was unacceptably large for t less than 25 seconds.

Fitting and testing the model. The model equations (5) were solved numerically, assuming an initial NAD concentration of 160 μM. The initial estimates of the parameters $k_1, \cdots, k_6$ were those of Theorell, Nygaard & Bonnichsen (1955). These values were successively modified until the calculated solution fitted the data within the limits of experimental error.

The validity of the model with these parameters was then checked by calculating solutions corresponding to the other data curves. Figure 1, which shows the duplicate sets of data and the numerical solutions, indicates that the model fits reasonably well for NAD concentrations between 80 μM and 240 μM.

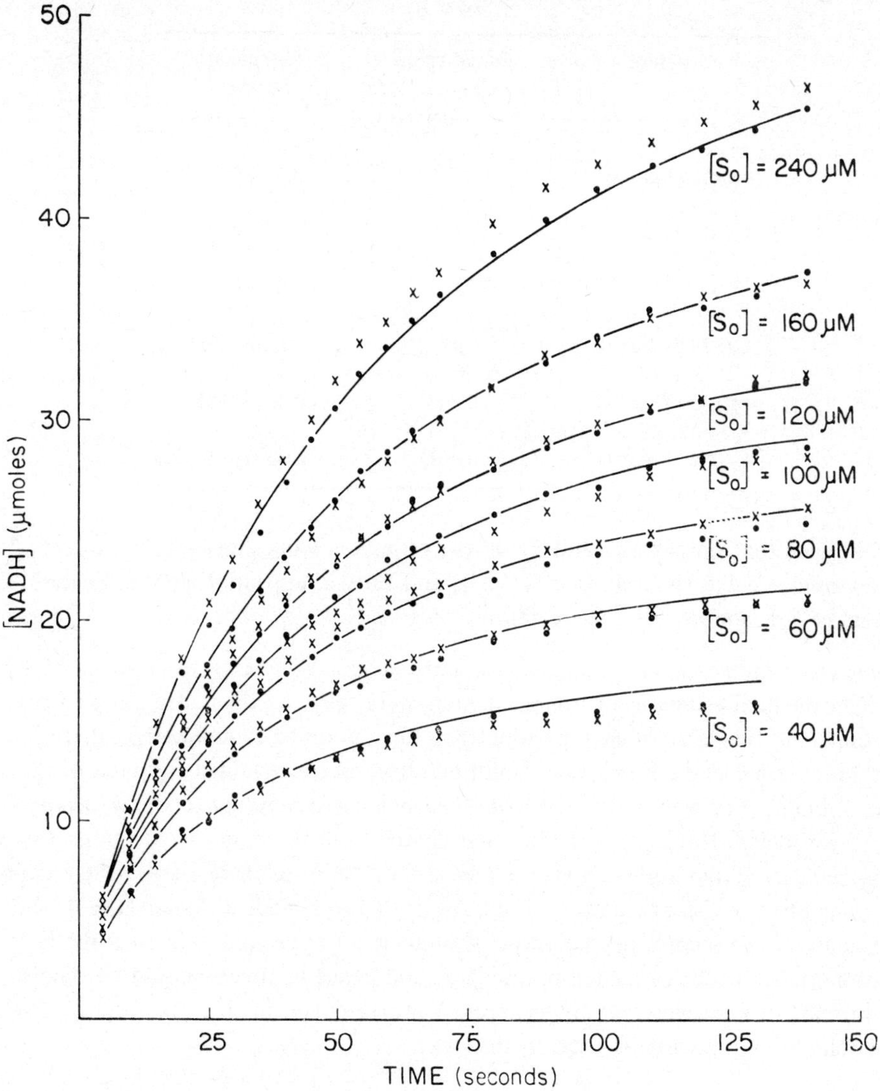

Fig. 1. Predicted values of NADH concentration $[E_0] = 0.63$ μM, (rate constants as in text).

——	predicted value
×	observed value, replicate I
·	observed value, replicate II

Confidence intervals for the parameters. Inspection of the data and consideration of the accuracy of measurement of the reactants indicated that experimental error could account for differences of 1 to 2 μM in the concentration of NADH. Therefore, for each parameter the change in value necessary to alter the calculated solution by 1 μM NADH was determined, holding the other values constant. The intervals thus constructed, together with the initial estimates used, are shown in Table 1.

Table 1

Constant	Initial value (Theorell *et al.*, 1955)	Estimated interval
$k_1(\mu M^{-1}\text{sec}^{-1})$	3.7	3.7 ± 0.5
$k_2(\text{sec}^{-1})$	1.6	2.1 ± 0.2
$k_3(\text{sec}^{-1})$	37.0	37.0 ± 8.0
$k_4(\mu M^{-1}\text{ sec}^{-1})$	0.24	0.23 ± 0.05
$k_5(\mu M^{-1}\text{ sec}^{-1})$	0.30	0.30 ± 0.05
$k_6(\mu M^{-1}\text{ sec}^{-1})$	0.0035	0.0070 ± 0.0005

It can be seen that only for k_2 and k_6 do the estimated values disagree to any extent with the values obtained by Theorell *et al.* (1955), and thus it appears that the Theorell-Chance mechanism is appropriate for this reaction.

Estimation of the variance matrix of the parameters. Following the method of Spendley *et al.* (1962), the final estimates of the parameters were taken as the coordinates of one vertex of a simplex, and the value of each parameter in turn modified to define the other six vertices in the 6-dimensional parameter space. Solution curves were obtained for each of these parameter combinations as well as for the combinations corresponding to the mid-points of the edges of the simplex. Deviations of the fitted curves from the data at the points $t = 35$ and $t = 70$ were squared and added to give a measure of the fit of these curves, then a quadratic function was fitted to these response values. According to Nelder & Mead (1965), the matrix of coefficients of the second degree terms should be an approximation to a multiple of the information matrix of the estimated parameters, and hence its inverse could be used to obtain their approximate variances and covariances. Unfortunately, in this case, the inverse is not positive definite and cannot be used in this way.

REFERENCES

MICHAELIS, L. & MENTEN, M. L. (1913). Die Kinetik der Invertinwirkung. *Biochem. Z.* **49,** 333–69.

NELDER, J. & MEAD, R. (1965). A simplex method for function minimization. *Computer J.* **7,** 308–13.

THEORELL, H. & BONNICHSEN, R. (1951). Studies on liver alcohol dehydrogenase I. Equilibria and initial reaction velocities. *Acta Chem. Scand.* **5,** 1105–26.

THEORELL, H. & CHANCE, B. (1951). Studies on liver alcohol dehydrogenase II. Kinetics of the compound of horse liver alcohol dehydrogenase and reduced diphosphopyridine nucleotide. *Acta Chem. Scand.* **5,** 1127–44.

THEORELL, H., NYGAARD, A. P. & BONNICHSEN, R. (1955). Studies on liver alcohol dehydrogenase III. The influence of pH and some anions on the reaction velocity constants. *Acta Chem. Scand.* **9,** 1148–65.

SPENDLEY, W., HEXT, G. R. & HIMSWORTH, F. R. (1962). Sequential application of simplex designs in optimization and evolutionary operation. *Technometrics* **4,** 441–61.

On assessing the progress of epidemics

NIELS BECKER
La Trobe University

1. Introduction

For the regulation of stochastic processes it is important to reassess our expectations of the process as more data become available. This presents no difficulties when the process is Markovian and our information on the present state of the process is complete. In practice one is often unable to obtain complete data and thus arises the need to determine the conditional process, given the available data. We consider this problem with reference to an epidemic process. For the control of communicable diseases it is important to assess the progress of an epidemic during the early stages of the outbreak. The data available for this purpose generally consist of a knowledge of (i) some properties of the disease, (ii) the number of cases observed, and (iii) the times between the detection of cases, i.e. the interremoval times. The number of infectives in the community and the times at which infections occur are usually not observable. Thus the state of the process is not determined by the data.

During the early stages of the outbreak some epidemic processes can be approximated by a branching process. By considering the first few interremoval times as observations on the approximating process, one can obtain recursive formulae for expressions relevant to the conditional process, given the observed interremoval times. These equations can be solved explicitly by taking specific offspring distributions. We also consider the estimation of the mean of the offspring distribution on the basis of the observed interremoval times.

2. The epidemic model

During the course of an epidemic an individual who is susceptible to the particular disease may be infected by some means such as breathing in some contagious pathogenic agent. This event usually initiates a complex internal process resulting in the elapse of a *latent period* during which there is no emission of infectious material. This is immediately followed by the *infectious period*, which continues until the infective is 'removed'.

The time points most relevant in an assessment of the progress of an epidemic are those at which infectives are detected and removed. It seems appropriate to formulate epidemic models with reference to these time points. Consider a disease for which the duration of the effective infectious period is short. During the early stages of an epidemic the reduction in the size of the susceptible population may be ignored and the growth of the infectious population may be approximated by a branching process; see Bharucha-Reid (1956), Neyman & Scott (1964), Bartoszynski (1967). Suppose that the duration of the latent period is exponentially distributed with parameter β, the removal rate for infectives. Let the probability generating function (p.g.f.) for the offspring distribution be $\phi(z)$ and let X_n denote the size of the infectious population immediately after the $(n+1)$th removal. The size of the infectious population just after the first removal is then X_0 which has p.g.f. $\phi(z)\chi(z)/z$, where $\chi(z)$ is he p.g.f. corresponding to the number of infectives who started the epidemic. The p.g.f. of X_{n+1} is given by

$$\phi_{n+1}(z) = \phi(z)\{\phi_n(z) - q_n\}/\{z(1-q_n)\} \qquad (n = 0, 1, 2, \cdots) \tag{1}$$

where $\phi_0(z) = \phi(z)\chi(z)/z$ and

$$q_n = \lim_{z\to 0} \phi_n(z) = \mathrm{pr}\,(X_n = 0 \mid X_{n-1} > 0). \tag{2}$$

The probability that the ultimate size of the epidemic is $n+1$ is given by

$$p_n = \mathrm{pr}\,(X_n = 0, X_{n-1} > 0) = q_n\left(1 - \sum_{i=0}^{n-1} p_i\right) \quad (n = 1, 2, \cdots)$$

with $p_0 = q_0$. A more direct way of computing these probabilities is to note that p_n is the coefficient of z^{n+1} in $\chi\{\Phi(z)\}$ where $\Phi(z)$ satisfies

$$\phi(z) = z\phi\{\Phi(z)\}$$

(Harris, 1963, p. 32). When $\chi(z) = z^r$, for some integer r, the p_n can be taken from the analogous distribution of the time-to-first-emptiness of a dam fed by independent, integer inputs, with a unit release. It follows (see, for example, Mott, 1963) that

$$p_n = \frac{r}{n+1} f(n-r+1, n+1)$$

where $f(n-r+1, n+1)$ is the probability that the total number of individuals infected by $n+1$ infectives is $n-r+1$ and is given by the coefficient of z^{n-r+1} in $\phi^{n+1}(z)$. For the case $\chi(z) = z^r$ and $\phi(z) = \exp(\mu z - \mu)$ this gives

$$p_n = r\mu e^{-(n+1)\mu}\{(n+1)\mu\}^{n-r}/(n-r+1)! \quad n = r-1, r, \cdots. \tag{3}$$

These probabilities are given in Table 1 for $r = 1$, $n = 0, 1, \cdots 5$ and various values of μ for later reference.

Table 1. *Values of p_n, the probability that the branching process terminates with $n+1$ cases, when the offspring distribution is Poisson with mean μ*

n \ μ	.25	.50	.75	1.00	1.50	2.00	2.50	3.00
0	.7788	.6065	.4724	.3679	.2231	.1353	.0821	.0498
1	.1516	.1839	.1673	.1353	.0747	.0366	.0168	.0074
2	.0443	.0837	.0889	.0747	.0375	.0149	.0052	.0017
3	.0153	.0451	.0560	.0488	.0223	.0072	.0019	.0004
4	.0058	.0267	.0388	.0351	.0145	.0038	.0008	.0001
5	.0024	.0168	.0285	.0268	.0101	.0021	.0003	.00004

The p.g.f.'s $\phi_n(z)$ are appropriate for assessing the progress of an epidemic when only the number of removals is observed. However, when $n+1$ removals have been observed we generally also have observations on T_k, the time between the kth and $(k+1)$th removal, $k = 1,2,\cdots,n$. We are therefore interested in the conditional p.g.f. of X_n, given the first n interremoval times. Dietz (1968) shows that the conditional generating functions

$$\psi_n(z;t_1,\cdots,t_n) = E(z^{X_n} | T_1 = t_1,\cdots,T_n = t_n)$$

satisfy the recursion formula

$$\psi_n(z;t_1,\cdots,t_n) = \phi(\ \psi'_{n-1}(ze^{-\beta t_n};t_1,\cdots,t_{n-1})/\psi'_{n-1}(e^{-\beta t_n};t_1,\cdots,t_{n-1})\quad (n=1,2,\cdots) \tag{4}$$

with $\psi_0(z) = \phi(z)\chi(z)/z$. The prime indicates the first partial derivative with respect to z. Unfortunately it does not seem possible to solve the recursive expression (4) without making further assumptions.

As an example suppose that the epidemic is started by one infective and that the offspring distribution is Poisson, with mean μ. Then $\chi(z) = z$ and $\phi(z) = \exp(\mu z - \mu)$, and we find

$$\psi_n(z;t_1,\cdots,t_n) = \exp\{\mu a_n(z-1)\},$$

where $a_0 = 1$ and $a_n = a_n(t_1,\cdots,t_n)$ is given by

$$a_n = 1 + a_{n-1}\exp(-\beta t_n) = 1 + \sum_{k=1}^{n} \exp\left(-\beta \sum_{j=k}^{n} t_j\right) \quad (n = 1,2,\cdots). \tag{5}$$

Our assumptions imply that X_n, the number of infectives immediately following the $(n+1)$th removal, given the n interremoval times $t_1,\cdots,t_n$, is a Poisson variable with mean μa_n.

3. The ultimate size of the epidemic

For a complete description of the ultimate size of an outbreak of a disease one needs to consider non-linear models. However, the model considered here can provide an adequate description of the progress of small outbreaks or the early stages of large outbreaks, and this is useful for deciding whether the first n observed interremoval times are indicative of a minor epidemic or a major epidemic.

If we accept the Threshold Theorem for Epidemics; see Bailey (1957), then either minor epidemics or major epidemics occur, with hardly any in between. This theorem may be deduced from branching processes by interpreting extinction of the process as a minor epidemic and non-extinction as a major epidemic; see Bartlett (1966), Bartoszynski (1967). As is well known, the probability that a Markov branching process ultimately becomes extinct, given a population size $X(t)$ at time t, is $q^{X(t)}$, where q satisfies $q = \phi(q)$. The probability of a minor epidemic can therefore be viewed as a stochastic process with moments which initially equal $\chi(q^r)$ and immediately following the $(n+1)$th removal equal $\phi_n(q^r)$ where q is the smallest non-negative root of $z = \phi(z)$. The conditional moments of q^X, given the first n interremoval times, are given by $\psi_n(q^r;t_1,\cdots,t_n)$, $(r = 1,2,\cdots)$. In the example, with one initial infective and a Poisson offspring distribution, the conditional probability of a minor epidemic, given $t_1,\cdots,t_n$ is q^{a_n}. This probability tends to decrease as more removals are observed, provided $q<1$.

The probability of a minor epidemic is unity whenever $\mu = \phi'(1) < 1$. In this case the model can provide an approximate description of the ultimate size of the epidemic. In particular, the mean of the ultimate size of the epidemic is $\chi'(1)/(1-\mu)$. The conditional mean, given $t_1,\cdots,t_n$, is $\psi_n'(1;t_1,\cdots,t_n)/(1-\mu)$. In the example $\chi(z)=z$ and $\phi(z)=\exp(\mu z-\mu)$, the conditional mean becomes $\mu a_n/(1-\mu)$.

4. Estimation of parameters

Inference based on the first few interremoval times becomes necessary when the epidemic model contains parameters whose values are likely to be different and unknown for each outbreak of the disease. It is perhaps questionable whether the first few interremoval times contain very much information, however, if the methods of analysis involve only simple formulae, which are easily updated as more data become available, then it is worth making use of this information. When the properties of the disease are well known and the standard control procedures have been decided upon, an adequate estimate of the removal rate for infectives might be available. The rate of infection, however, is community dependent and is therefore likely to be different and unknown for each outbreak of the disease. We now take the above epidemic model with $\chi(z)=z$ and $\phi(z)=\exp(\mu z-\mu)$ and, by considering the estimation of μ, use it to illustrate some problems inherent in estimation for processes which may be terminating or non-terminating.

For the purpose of estimation it is not practical to wait until the process terminates because the process considered here might not terminate and because the model does not provide an adequate description of the later stages of large epidemics. Further, if the information is to be used for deciding on an appropriate control measure for the epidemic, then there is no sense in waiting until the epidemic is over. It is not wise to set aside a fixed time interval for taking observations since it is possible, although perhaps unlikely, that a very large number of cases are observed in the time interval. It is safer to decide on a stopping rule based on the number of removals. We therefore consider the truncated estimator based on $t_1,\cdots,t_m$ and m if the process terminates with $m+1$ cases ($m<n$) and based on $t_1,\cdots,t_n$ and n if the process has not terminated by then.

The conditional density function of the interremoval time T_k, given $T_1=t_1,\cdots,T_{k-1}=t_{k-1}$ and $X_{k-1}=x$, is $x\beta\exp(-x\beta t)$, $t>0$. It follows that the conditional density function for T_k, given $t_1,\cdots,t_{k-1}$, is given by

$$g_k(t\,|\,t_1,\cdots,t_{k-1}) = \beta e^{-\beta t}\psi_{k-1}'(e^{-\beta t};t_1,\cdots,t_{k-1}).$$

The likelihood of the observed intervals $t_1,\cdots,t_n$ is thus given by

$$\begin{aligned} f(t_1,\cdots,t_n) &= \prod_{k=1}^{n} g_k(t_k\,|\,t_1,\cdots,t_{k-1}) \\ &= \beta^n \exp\left(-\beta\sum_{k=1}^{n} t_k\right)\prod_{k=1}^{n}\psi_{k-1}'(e^{-\beta t_k};t_1,\cdots,t_{k-1}). \end{aligned}$$

When there is one initial infective and a Poisson offspring distribution, the likelihood function simplifies to

$$f(t_1,\cdots,t_n) = \mu^n\beta^n\left(\prod_{k=1}^{n-1} a_k\right)\exp\left\{-\beta\sum_{k=1}^{n} t_k-\mu(n+1-a_n)\right\}. \tag{6}$$

It is seen that $a_n(t_1,\cdots,t_n)$ is sufficient for μ and that the maximum likelihood estimator of μ at the time of the $(n+1)$th removal is

$$M_n^+ = n/(n+1-A_n),$$

where $A_n = a_n(T_1, \cdots, T_n)$. The estimates of μ are easily updated as more information becomes available, by using (5). Note that M_n^+ is a conditional estimator, given that the ultimate size of the epidemic exceeds n, which suggests that M_n^+ might tend to overestimate the value of μ. It seems possible that this tendency, which was indeed observed in simulations of the process, will be countered when the possible termination of the epidemic prior to $n+1$ removals is allowed for. If there are a total of $m+1$ cases, where $m < n$, we will in practice observe interremoval times $t_1, \cdots, t_m$ and the event $T_{m+1} > \tau$, where τ is a fixed constant. The corresponding likelihood, namely

$$\mu^m \beta^m \left(\prod_{k=1}^{m-1} a_k\right) \exp\left\{-\beta \sum_{k=1}^{m} t_k - \mu(m+1-a_m e^{-\beta\tau})\right\} \tag{7}$$

is obtained by using

$$\mathrm{pr}(T_{m+1} > \tau \mid t_1, \cdots, t_m) = \psi_m(e^{-\beta\tau};\ t_1, \cdots, t_m).$$

When τ is large the likelihood takes the form

$$\mu^m e^{-(m+1)\mu} h(t_1, \cdots, t_m).$$

Hence m is sufficient for μ in this case and the corresponding maximum likelihood estimate is $m/(m+1)$.

The combined estimator M_n is then partially a discrete random variable and partially a continuous random variable. The discrete part is approximated by

$$\mathrm{pr}\{M_n = m/(m+1)\} = p_m \qquad (m = 0, 1, \cdots, n-1)$$

where the p_m is given by (3) with $r = 1$. The p_m are given for various values of μ and $m = 0, 1, \cdots, 5$ in Table 1. The continuous part is best described by making a change of variable in (6) from $t_1, \cdots, t_n$ to $a_1, \cdots, a_{n-1}$, $m_n = n/(n+1-a_n)$, which gives the joint 'density' function

$$n\mu^n m_n^{-2} e^{-n\mu/m_n}, \left\{1 \leqq a_k \leqq 1 + a_{k-1},\ (k = 1, \cdots, n-1),\ 1 \leqq m_n \leqq \frac{n}{n - a_{n-1}}\right\}.$$

The integral of this function over the feasible region for $a_1, \cdots, a_{n-1}$ and m_n, gives $1 - \sum_{k=0}^{n-1} p_k$, the probability of at least $n+1$ cases.

It does not seem possible to express the moments of M_n in terms of elementary functions, which suggests that M_n is generally biased for μ. In particular $E(M_1)$ is infinite, although $\mathrm{pr}(M_1 = 0)$ can be arbitrarily close to unity. When the ultimate size of the epidemic is less that $n+1$ the estimate for μ is less than unity and otherwise it is greater. In view of the discrete nature of M_n in $[0, 1]$, the estimate of μ cannot generally be arbitrarily close to μ. When $\mu = 0.25$, for example, the estimate necessarily differs from μ by at least 0.25.

Note also that in practical terms it is meaningless to have $\mu = 0$, yet the estimator is zero with probability $e^{-\mu}$ which is large for $\mu < 1$ and even for $\mu > 1$ can exceed 0.35. Given that a process is going to terminate it is likely to do so before many infections have taken place, both when μ is large and when μ is small, as can be seen from the probabilities given in Table 1. It seems, therefore, that when the process terminates with a small number of cases, the estimator is not very reliable.

The fact that m is sufficient for μ, when the epidemic terminates with $m + 1$ cases, implies that $t_1, \cdots, t_m$ provide no further information on μ. It is difficult to establish analytically how much information is contained in our estimator, since the $t_1, \cdots, t_n$ are not independent and the usual asymptotic properties of maximum likelihood estimators do not apply. In particular, we note that the estimator is clearly not consistent, due to the possibility of early termination of the process. Under the circumstances it is useful to simulate the process, and observe how the estimator performs empirically.

Table 2 summarizes the results of 1000 realizations of the branching process with $\mu = 1$. It is not necessary to choose a value for β since the expression for the estimator involves β only in the form $e^{-\beta T}$, where T is an exponential variate, which is of the form $U^{1/x}$ where U is distributed uniformly on $(0, 1)$. The m_n were computed sequentially, using (5), from the same 1000 realizations, and thus $m_1, m_2, \cdots$ were not generated independently. This procedure clearly saves considerably on computer time. The values that the estimator can assume for terminating realizations are given in column two. The proportion of realizations terminating with $n + 1$ cases is given in column three. A comparison of these proportions with the corresponding probabilities of Table 1 provides some assurance that the simulated realizations are 'reasonably typical'. The information provided by the simulation on non-terminating realizations is summarized by the sample means and standard deviation of the estimates m_n^+ in columns five and six. Apart from providing orders of magnitudes for these, the simulations

Table 2. *Sample means and standard deviations of m_n based on* 1000 *simulated realizations of the branching process when the offspring distribution is Poisson with mean* $\mu = 1$

n	Terminating realizations		Non-terminating realizations			All realizations	
	$\frac{n-1}{n}$	Proportion of realizations with n cases	Number of realizations exceeding n cases	Average of the m_n^+	Standard deviation of the m_n^+	Average of the m_n	Standard deviation of the m_n
1	.0000	.384	616	6.638	15.986	4.089	12.952
2	.5000	.096	520	2.962	3.001	1.588	2.597
3	.6667	.057	463	2.168	1.553	1.090	1.468
4	.7500	.059	404	1.963	.989	.923	1.087
5	.8000	.045	359	1.824	.744	.823	.913
6	.8333	.014	345	1.650	.553	.747	.780
7	.8571	.025	320	1.624	.512	.719	.742
8	.8750	.027	293	1.564	.419	.681	.681
9	.8889	.020	273	1.532	.402	.659	.653
10	.9000	.012	261	1.531	.385	.651	.642
11	.9091	.010	251	1.513	.363	.640	.626
12	.9167	.003	248	1.471	.383	.628	.612
13	.9231	.019	229	1.494	.379	.623	.607
14	.9286	.004	225	1.468	.340	.615	.591
15	.9333	.005	220	1.432	.311	.604	.572
16	.9375	.003	217	1.390	.300	.594	.556
17	.9412	.008	209	1.385	.267	.589	.547
18	.9444	.010	199	1.390	.250	.586	.541
19	.9474	.000	199	1.373	.241	.582	.536
20	.9500	.005	194	1.364	.235	.578	.530

suggest that, as n increases, the sample mean approaches unity from above while the standard deviation decreases. Similar results were observed for simulations carried out with values of $\mu < 1$. For values of $\mu > 1$ the sample means were observed to approach the true value of μ from above, while the standard deviations decreased with n. The pooled sample means and standard deviation, i.e. for all realizations, are given in columns seven and eight of Table 2, for $\mu = 1$. The pooled means decrease monotonically to a value well below the value for μ and the standard deviations tend to remain quite large. Simulations with other values for μ produced similar results. It seems possible to decrease the bias of the estimator by adjusting the zero value it takes when only one removal is observed. Attempts at this were not uniformly successful for all values of μ.

One point requiring further comment concerns the choice of τ, the time one ought to wait before deciding that the epidemic has been terminated. The decision is to observe n inter-removal times. As long as there are less than $n+1$ cases we should be prepared to wait, provided only that the assumptions of the model remain valid. Consequently, when we are considering a disease of which there are infrequent outbreaks, we should be prepared to wait for a long time, before deciding that the epidemic has terminated. When considering a disease of which there are more frequent epidemics, we cannot wait too long since a further case following a long wait could well be the first of a new epidemic. Since $1 < a_m \leqq m+1$, the term in (7) containing τ may be ignored if $\exp(-\beta\tau)$ is small compared with unity. Therefore the probability that the length of the latent period exceeds τ should be small. Taking

$$\tau > 4 \times \text{mean duration of the latent period}$$

implies $\exp(-\beta\tau) < .0184$, which should be adequate for most purposes.

There are two important questions on whose solution we have merely made vague hints. The first is whether M_n^+ is a consistent estimator of μ. More generally one may ask under what conditions, for a process which can be terminating or non-terminating, there exists a conditional estimator which is consistent for a parameter of the process. That conditional estimators which are consistent do exist for processes with absorbing states is known (see, for example, Matthews, 1970). The second question is: How much information is contained in the interremoval times? The dependence of the interremoval times encourages one to ignore most of this information; see Bailey (1957). It clearly becomes important to determine how much information is lost due to this action.

I am indebted to the referee for referring me to the result in dam theory, which is useful in determining the p_n when the initial number of infectives is fixed.

References

BAILEY, N. T. J. (1957). *The Mathematical Theory of Epidemics.* London: Griffin; New York: Hafner.

BARTLETT, M. S. (1966). *Stochastic Processes.* 2nd edition. Cambridge University Press.

BARTOSZYNSKI, R. (1967). Branching processes and the theory of epidemics. *Proc. 5th Berkeley Symp.*, **4**, 259–69.

BHARUCHA-REID, A. T. (1956). On the stochastic theory of epidemics. *Proc. 3rd Berkeley Symp.*, **4**, 111–9.

DIETZ, K. (1968). Some quickest detection problems for terminating point processes. Center for Environmental Quality Management. Cornell University. Report no. 1008.

HARRIS, T. E. (1963). *The Theory of Branching Processes.* Berlin: Springer-Verlag.

MATTHEWS, J. P. (1970). A central limit theorem for absorbing Markov chains. *Biometrika*, **57**, 129–39.

MOTT, J. L. (1963). The distribution of the time-to-emptiness of a discrete dam under steady demand. *J. R. Statist. Soc. B*, **25**, 137–9.

NEYMAN, J. & SCOTT, E. (1964). A stochastic model of epidemics. In *Stochastic Models in Medicine and Biology*, ed. J. Gurland, 45–85. University of Wisconsin Press.

Statistical estimation in elasticity theory

J. J. CROSS
University of Melbourne

1. Introduction

In the literature on theoretical mechanics much attention is paid to the derivation of exact nonlinear theories of stress and strain. The verification of these theories is left to the experimentalists, who reduce the general theories and formulae to particular cases and situations, measure several quantities, do some calculations, and then make claims. When questioned about the "reproducibility" of their data, some experimenters blithely say that it is "very good." The fact that different figures result when the same experiment is repeated several times does not seem to matter, does not seem to shake their faith in the deterministic nature of mechanics, except that some "averaging" must be done.

The present article stems from several such contacts with experimenters. In section 2 we analyse the simplest elastic situation, and estimate the extension of a rod under uniaxial tension or compression. Section 3 presents the general three-dimensional elastic situation, while §§4 and 5 provide estimators for two matrices fundamental to elasticity theory. No attempt is made here towards the estimation of the physical constants or functions which appear in the stress-strain relations.

2. Uniaxial extension

Consider uniaxial extension of a homogeneous, isotropic, elastic body in the form of a rod (Fig. 1). The stress produced by the extension is assumed to follow a linear Hooke's law from theoretical considerations:

$$\sigma \equiv E\varepsilon, \tag{1}$$

where σ is the stress (force per unit area in the strained state), and ε is the extension, with both σ and ε being along the Ox axis; E is a constant dependent on the material only. The extension ε is defined as

$$\varepsilon \equiv \frac{dy}{dx} - 1, \tag{2}$$

where x is the position of the material point along the Ox axis before the extension, and $y = y(x)$ its position after the deformation.

In experimental work and in some theoretical situations, it is convenient to replace the derivative dy/dx by the ratio of the small increments whose limit the derivative is. If we write $v \equiv y(x + u) - y(x)$ for the increment in y corresponding to the increment u in x, we may rewrite (2) in the form:

$$\varepsilon = \frac{v}{u} - 1 = \frac{v - u}{u}; \tag{3}$$

i.e., the extension ε is the ratio of the difference $v - u$ of the lengths of a small segment at the point x before (u) and after (v) the deformation, to the undeformed length (u).

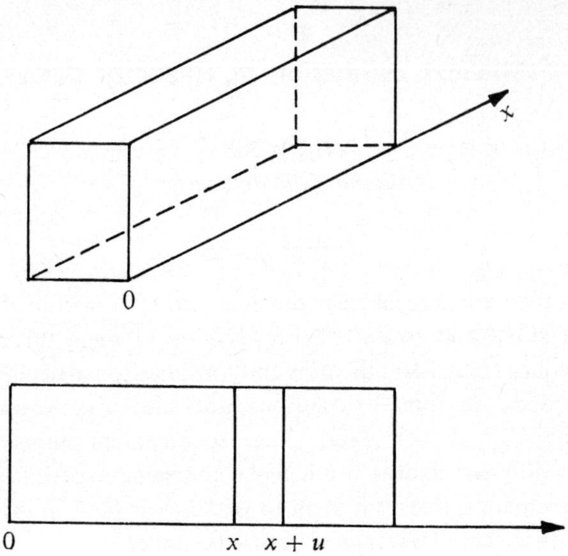

(*a*) Before deformation.

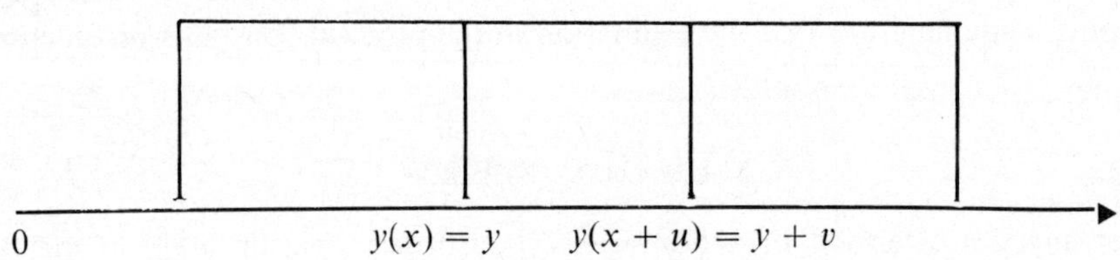

(*b*) After deformation.

Figure 1.

Several problems may be considered:

(i) the Young's modulus E may be known and one of σ and ε measured, and (1) then used to calculate the other; (ii) both σ and ε may be measured, and (1) used to calculate the constant E.

There is usually some error in the measurement of the stress σ or the strain ε, so that some "averaging" procedure must be employed. It is a feature of the experimental literature, that little or no attention is paid to the statistical and probabilistic nature of these measurements, or to the validity of the "averaging" procedure used. In fact, experimenters often fail to reveal the "averaging" procedures they have employed to calculate their values of the Young's modulus E, as well as other moduli and formulae. It would be invidious to single out any individual experimenter or journal, but ten minutes among the experimental literature on the shelves of any Engineering library will show the truth of the above statements.

Basic to almost every use of (1) is an estimation of ε. The extension ε is measured in many ways: by resistance strain gauges as described in Perry & Lissner (1962), by photoelasticity as described in Dally & Riley (1965), by semiconductor strain gauges as described in Sanchez & Wright (1962), or by grid patterns scratched on the material and copied by a soft-metal technique as described in Hickson (1965). The resistance and semiconductor strain gauges measure the extension, in one or several directions, in a quick and handy fashion, but they are rough and inaccurate; their error is large and hard to assess, and they give ε directly, the local lengths u and v not being measured. The soft-metal grid replica technique of Hickson, while perhaps laborious, is more accurate and allows the error in the value of the extension to be estimated and confidence limits for the value to be stated. This we now show.

The grid replica technique copies a pattern of scratches before and after the extension, and enables us to make several measurements of u and v near each point where we wish to find ε; so ε is to be estimated from a sample of, say, n paired measurements:

$$(u_1, v_1), (u_2, v_2), \cdots, (u_n, v_n).$$

Rewrite (3) as

$$v = (1 + \varepsilon)u.$$

Now regard the unstrained displacement u as being known exactly, i.e., regard u as a mathematical variable, and take v as being subject to a random error e. We may now employ the linear regression technique to estimate the coefficient $(1 + \varepsilon)$ from the assumed relationship:

$$v_\alpha = (1 + \varepsilon)u_\alpha + e_\alpha, \quad \alpha = 1, 2, \cdots, n.$$

If we adopt the usual least-squares estimator, we get

$$1 + \hat{\varepsilon} = \sum_{\alpha=1}^{n} v_\alpha u_\alpha \Big/ \sum_{\alpha=1}^{n} u_\alpha^2 .$$

Thus ε is estimated by $\hat{\varepsilon}$:

$$\hat{\varepsilon} = \sum_{\alpha=1}^{n} (u_\alpha[v_\alpha - u_\alpha]) \Big/ \sum_{\alpha=1}^{n} u_\alpha^2 .$$

If each e_α has the distribution $N(0, \tau^2)$, i.e., if each is normally distributed with zero mean and variance $\tau^2 > 0$, then each v_α has distribution $N((1 + \varepsilon)u_\alpha, \tau^2)$, and the estimator $\hat{\varepsilon}$ has the distribution $N(\varepsilon, \tau^2 / \sum_{\alpha=1}^{n} u_\alpha^2$. The variance τ^2 may be estimated by $\hat{\tau}^2$:

$$(n-1)\hat{\tau}^2 = \sum_{\alpha=1}^{n} [v_\alpha - (1 + \hat{\varepsilon})u_\alpha]^2 \tag{4}$$

and $(n-1)\hat{\tau}^2/\tau^2$ has the χ^2_{n-1} distribution, i.e., a central chi-square distribution with $(n-1)$ degrees of freedom. A better form of (4) is

$$(n-1)\hat{\tau}^2 = \sum_{\alpha=1}^{n} v_\alpha^2 - (1 + \hat{\varepsilon})^2 \sum_{\alpha=1}^{n} u_\alpha^2 .$$

The purpose of this paper is to generalize the estimation process to more complete strain measures, and to estimate their errors, if practicable. This is taken up after the next section, in which we present some elasticity theory in order to see which strain measures need estimation.

3. ELASTICITY THEORY

The basic situation in the ordinary continuum theory of elasticity is as follows. The experimental or theoretical body $\mathscr{B}$ occupies some region $\mathscr{R}$ of three-dimensional euclidean space at time $t = 0$, say. This region or position of the body is called the undeformed or unstrained state of the body. This state is referred to rectangular cartesian axes fixed in space. The coordinates of a typical point p of the body $\mathscr{B}$ in the undeformed state are taken to be (x_1, x_2, x_3). If we regard the euclidean space as having its origin fixed at the coordinate origin, we may regard the position of p as a vector and write

$$\boldsymbol{x} = \sum_{k=1}^{3} x_k \boldsymbol{b}_k$$

where the three vectors $\boldsymbol{b}_1$, $\boldsymbol{b}_2$, $\boldsymbol{b}_3$ are unit vectors along the appropriate coordinate axes (Fig. 2).

After the deformation, the body will occupy some region $\bar{\mathscr{R}}$, and the same material point p will have coordinates labelled (y_1, y_2, y_3), and we may write

$$\boldsymbol{y} = \sum_{k=1}^{3} y_k \boldsymbol{b}_k .$$

For continuous deformations (no cracks or tears), it is usual to assume that $\boldsymbol{y}$ is a smooth, invertible function of $\boldsymbol{x}$ over the whole body $\mathscr{B}$, i.e., that

$$\boldsymbol{y} = \boldsymbol{y}(\boldsymbol{x})$$

or

$$y_i = y_i(x_1, x_2, x_3), \qquad i = 1, 2, 3$$

are smooth functions, and that

$$\det\left(\partial y_i / \partial x_k\right) \neq 0$$

everywhere in $\mathscr{B}$ (Truesdell & Noll, 1965). In short, the function $\boldsymbol{y}$ is a diffeomorphism on $\mathscr{B}$.

Further, it is commonly assumed that the functions $\boldsymbol{y}$ or y_i defining the deformation can be expanded locally, about each material point p, as Taylor's series in several variables; i.e., if p' is a material point near p, and if p' has position vectors $\boldsymbol{x}'$, $\boldsymbol{y}'$ in the undeformed and deformed states of $\mathscr{B}$, respectively, then

$$\boldsymbol{y}' = \boldsymbol{y} + \boldsymbol{F}(\boldsymbol{x}' - \boldsymbol{x}) + O(|\boldsymbol{x}' - \boldsymbol{x}|^2),$$

or

$$y_i' = y_i + \sum_{j=1}^{3} F_{ij}(x_j' - x_j) + O\left(\sum_{k=1}^{3} (x_k' - x_k)^2\right),$$

where $F_{ij} = \partial y_i / \partial x_j$ evaluated at p. If we adopt the notation $\boldsymbol{y}' - \boldsymbol{y} = \boldsymbol{v}$, $\boldsymbol{x}' - \boldsymbol{x} = \boldsymbol{u}$, then we can rewrite the expansion as

$$\boldsymbol{v} = \boldsymbol{F}\boldsymbol{u} + O(|\boldsymbol{u}|^2) \tag{5}$$

or

$$v_i = \sum_{j=1}^{3} F_{ij} u_j + O(u_1^2 + u_2^2 + u_3^2). \tag{5'}$$

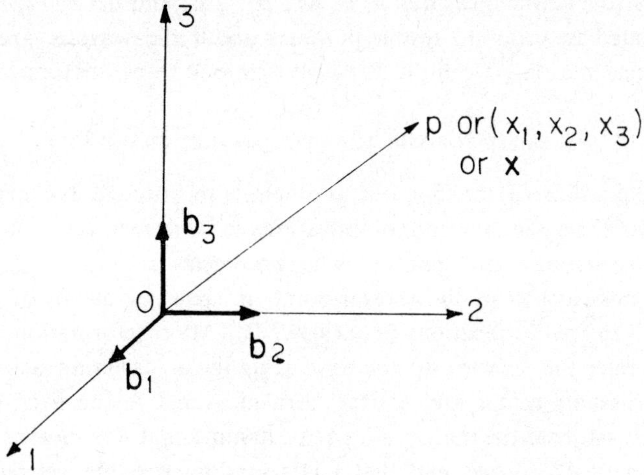

Figure 2

The matrix $\boldsymbol{F}$ called the deformation gradient; it is a function of $\boldsymbol{x}$ and is independent of $\boldsymbol{u}$.

In simple elasticity theories, the stress $\boldsymbol{T}$ is taken as a function of the deformation gradient $\boldsymbol{F}$, i.e.,

$$\boldsymbol{T} = \boldsymbol{H}(\boldsymbol{F})$$

or

$$T_{ij} = H_{ij}(F_{km}), \qquad (i,j,k,m = 1,2,3)$$

where each H_{ij} is a function of all nine components of $\boldsymbol{F}$. The stress matrix $\boldsymbol{T}$ and the function $\boldsymbol{H}$ may be subjected to further restrictions, such as invariance under certain changes in the undeformed state, invariance under rotations of the axes, and linearization. Some of the more common forms of $\boldsymbol{T}$ and $\boldsymbol{H}$ may be summarized as follows:

(i) For an elastic, isotropic, solid body,

$$\boldsymbol{T} = \phi_0(I_{\boldsymbol{B}}, II_{\boldsymbol{B}}, III_{\boldsymbol{B}})\boldsymbol{I} + \phi_1(I_{\boldsymbol{B}}, II_{\boldsymbol{B}}, III_{\boldsymbol{B}})\boldsymbol{B} + \phi_2(I_{\boldsymbol{B}}, II_{\boldsymbol{B}}, III_{\boldsymbol{B}})\boldsymbol{B}^2, \tag{6}$$

where $\boldsymbol{B} = \boldsymbol{F}\boldsymbol{F}^T$; $\boldsymbol{I}$ is the 3×3 identity matrix; $I_{\boldsymbol{B}}$, $II_{\boldsymbol{B}}$ and $III_{\boldsymbol{B}}$ are the three principal invariants of $\boldsymbol{B}$; and ϕ_0, ϕ_1 and ϕ_2 are scalar functions of these invariants; if the body is fluid, then (6) reduces to

$$\boldsymbol{T} = -p(\rho)\boldsymbol{I},$$

where p is a scalar function of the density $\rho = \rho_0 |\det \boldsymbol{F}|$, ρ_0 being a constant (Truesdell & Noll, 1965).

(ii) For an elastic, isotropic, incompressible, rubber body,

$$\boldsymbol{T} = -p\boldsymbol{I} + \mu(\tfrac{1}{2} + \beta)\boldsymbol{B} - \mu(\tfrac{1}{2} - \beta)\boldsymbol{B}^{-1},$$

where p is an arbitrary scalar, and μ and β are material constants (Truesdell & Noll, 1965).

(iii) For a linear elastic, isotropic body,

$$\boldsymbol{T} = \lambda(\operatorname{tr} \boldsymbol{E})\boldsymbol{I} + 2\mu\boldsymbol{E}, \tag{7}$$

where $2\boldsymbol{E} = \boldsymbol{F} + \boldsymbol{F}^T - 2\boldsymbol{I}$, and λ and μ are material constants (Sokolnikoff, 1956). Equation (1) is a particular case of (7) for a very special deformation.

Thus we see that the strain measures $\boldsymbol{F}$, $\boldsymbol{B}$, $\boldsymbol{B}^2$, $\boldsymbol{B}^{-1}$, $\boldsymbol{E}$, and $\det \boldsymbol{F}$, I_B, II_B and III_B would have to be estimated in order to test hypotheses about the various stress functions listed above. For practical reasons, we limit our investigations to estimators for $\boldsymbol{F}$, $\boldsymbol{B}$ and $\boldsymbol{E}$.

4. An estimator for the deformation gradient $\boldsymbol{F}$

If we return to equations (5) or (5′), our problem is to estimate the matrix $\boldsymbol{F}$ from measurements of the variables $\boldsymbol{v}, \boldsymbol{u}$. Several complications arise immediately: it is reasonably easy to measure $\boldsymbol{u}$ by constructing a test specimen whose geometry is simple, e.g.,a rectangular rod; but it is by no means easy to fix the several points in space by means of which the grids or grid replicas are given spatial locations (Turkin, 1972). After deformation, it may be a quite difficult matter to refer the vector $\boldsymbol{v}$ to the fixed rectangular cartesian axes.

However, if we assume that $\boldsymbol{v}$ and $\boldsymbol{u}$ have been measured in the fixed coordinate frame, we may proceed to estimate the matrix $\boldsymbol{F}$. Let us assume that $\boldsymbol{u}$ is measured accurately, i.e., that $\boldsymbol{u}$ is a mathematical variable, and that $\boldsymbol{v}$ is a random variable, related to $\boldsymbol{u}$ by

$$\boldsymbol{v} = \boldsymbol{F}\boldsymbol{u} + \boldsymbol{e},$$

where $\boldsymbol{v}$ is a 3×1 column of random variables, $\boldsymbol{F}$ a 3×3 matrix of constants, $\boldsymbol{u}$ a 3×1 column which is known exactly, and $\boldsymbol{e}$ a 3×1 column vector of random variables with mean $\boldsymbol{O}$ and 3×3 covariance matrix Σ.

Define several new matrices from a sample $(\boldsymbol{u}_1, \boldsymbol{v}_1)$, $(\boldsymbol{u}_2, \boldsymbol{v}_2), \cdots, (\boldsymbol{u}_n, \boldsymbol{v}_n)$, or $\{(\boldsymbol{u}_\alpha, \boldsymbol{v}_\alpha), \alpha = 1, 2, \cdots, n\}$, as follows:

$$\boldsymbol{X} = [\boldsymbol{u}_1 | \boldsymbol{u}_2 | \cdots | \boldsymbol{u}_n] = \begin{bmatrix} u_{11} & u_{12} & \cdots & u_{1n} \\ u_{21} & u_{22} & \cdots & u_{2n} \\ u_{31} & u_{32} & \cdots & u_{3n} \end{bmatrix},$$

$$\boldsymbol{A} = \boldsymbol{X}\boldsymbol{X}^T = \boldsymbol{A}^T, \tag{8}$$

$$\boldsymbol{Y} = [\boldsymbol{v}_1 | \boldsymbol{v}_2 | \cdots | \boldsymbol{v}_n] = \begin{bmatrix} v_{11} & v_{12} & \cdots & v_{1n} \\ v_{21} & v_{22} & \cdots & v_{2n} \\ v_{31} & v_{32} & \cdots & v_{3n} \end{bmatrix},$$

and

$$\boldsymbol{C} = \boldsymbol{Y}\boldsymbol{X}^T,$$

where $\boldsymbol{X}$ is a $3 \times n$ matrix, $\boldsymbol{A}$ 3×3, $\boldsymbol{Y}$ $3 \times n$, and $\boldsymbol{C}$ 3×3; the components of $\boldsymbol{X}$ are $u_{i\alpha}$, those of $\boldsymbol{Y}$ are $v_{i\alpha}$. The set of vectors $\{\boldsymbol{u}_1\, \boldsymbol{u}_2, \cdots, \boldsymbol{u}_n\}$ is so chosen that at least one triple of vectors forms a linearly independent set; hence, $\operatorname{rank}(\boldsymbol{X}) = 3$, $\operatorname{rank}(\boldsymbol{A}) = 3$ and $\boldsymbol{A}$ is invertible.

If we now apply least squares in order to minimize the total error:

$$\sum_{\alpha=1}^{n} (\boldsymbol{v}_\alpha - \boldsymbol{F}\boldsymbol{u}_\alpha)^T (\boldsymbol{v}_\alpha - \boldsymbol{F}\boldsymbol{u}_\alpha)$$

we obtain the normal equations

$$C - \hat{F}A = O$$

which yield

$$\hat{F} = CA^{-1} = YX^T(XX^T)^{-1},$$

since A is invertible. The matrix $\hat{F}$ is the least-squares estimator for F, and if we assume that the v_α have the multivariate normal distribution $N(Fu_\alpha, \Sigma)$, then $\hat{F}$ is also the maximum likelihood estimator for F. The maximum likelihood estimator $\hat{\Sigma}$ for Σ is

$$n\hat{\Sigma} = YY^T - \hat{F}A\hat{F}^T, \tag{9}$$

as in Anderson (1958).

The distributions of $\hat{F}$ and $\hat{\Sigma}$ are known when the v_α are independently distributed as $N(Fu_\alpha, \Sigma)$. Following Anderson, we see that

$$\begin{aligned} \mathscr{E}(\hat{F}) &= \mathscr{E}(YX^T(XX^T)^{-1}) \\ &= (\mathscr{E}Y)X^T(XX^T)^{-1} \\ &= FXX^T(XX^T)^{-1} = F. \end{aligned}$$

Further, the covariance matrix for the elements of $\hat{F}$ can be obtained from

$$\begin{aligned} &\mathscr{E}(\hat{F}_{ik} - F_{ik})(\hat{F}_{jm} - F_{jm}) \\ &= \mathscr{E}(v_{i\alpha}u_{r\alpha}A^{-1}_{rk} - (\mathscr{E}v_{i\alpha})u_{r\alpha}A^{-1}_{rk})j(v_{j\beta}u_{s\beta}A^{-1}_{sm} - (\mathscr{E}v_{j\beta})u_{s\beta}A^{-1}_{sm}) \\ &= \{\mathscr{E}(v_{i\alpha} - \mathscr{E}v_{i\alpha})(v_{j\beta} - \mathscr{E}v_{j\beta})\}\, u_{r\alpha}u_{s\beta}A^{-1}_{rk}A^{-1}_{sm} \\ &= \delta_{\alpha\beta}\, \Sigma_{ij}u_{r\alpha}u_{s\beta}A^{-1}_{rk}A^{-1}_{sm}, \text{ from independence,} \\ &= \Sigma_{ij}A_{rs}A^{-1}_{rk}A^{-1}_{sm}, \text{ by (3.3)} \\ &= \Sigma_{ij}A^{-1}_{km}. \end{aligned} \tag{10}$$

So $\hat{F}$ has a multivariate normal distribution with mean F, and covariance matrix with components as shown, and $n\hat{\Sigma}$ has a Wishart distribution, $W(\Sigma, n-3)$, from Anderson (1958, Ch. 8).

We define two matrix products for convenience. The left direct product $A \times B \equiv C$ is defined for square matrices A and B, and C has components

$$C_{rs} = A_{ij}B_{km},$$

where

$$\begin{aligned} r &= p(k-1) + i,\ 1 \leqq i \leqq p,\ 1 \leqq k \leqq q, \\ s &= p(m-1) + j,\ 1 \leqq j \leqq p,\ 1 \leqq m \leqq q, \end{aligned}$$

and A is of order p, B of order q, and C is of order pq.

The covariance product $A * B \equiv D$ is defined for square matrices A and B of order q, and D has components

$$D_{rs} = A_{ik}B_{jm}$$

where

$$\begin{aligned} r &= q(k-1) + i,\ 1 \leqq i, k \leqq q, \\ s &= q(m-1) + j,\ 1 \leqq j, m \leqq q, \end{aligned}$$

so that $\boldsymbol{D}$ is of order q^2.

We now see that

$$\mathscr{E}(\hat{\boldsymbol{F}} - \boldsymbol{F}) * (\hat{\boldsymbol{F}} - \boldsymbol{F}) = \boldsymbol{\Sigma} \times \boldsymbol{A}^{-1},$$

and that $\hat{\boldsymbol{F}}$ has distribution $N(\boldsymbol{F}, \boldsymbol{\Sigma X A}^{-1})$. In this notation, the matrix $\boldsymbol{F}$ is regarded as a vector whose components are $(F_{11}, F_{21}, F_{31}, F_{12}, F_{22}, F_{32}, F_{13}, F_{23}, F_{33})$, and we shall return to this in the next section.

We note that the multivariate distributions for $\hat{\boldsymbol{F}}$ and $\hat{\boldsymbol{\Sigma}}$ parallel the univariate distributions of $1 + \hat{\varepsilon}$ and $\hat{\tau}^2$ of §2.

It is now easy to find an estimator for $\boldsymbol{E} = \frac{1}{2}(\boldsymbol{F} + \boldsymbol{F}^T - 2\boldsymbol{I})$. If we put

$$\hat{\boldsymbol{E}} = \tfrac{1}{2}(\hat{\boldsymbol{F}} + \hat{\boldsymbol{F}}^T - 2\boldsymbol{I});$$

then

$$\mathscr{E}(\hat{\boldsymbol{E}}) = \boldsymbol{E},$$

and

$$\mathscr{E}(\hat{\boldsymbol{E}} - \boldsymbol{E}) * (\hat{\boldsymbol{E}} - \boldsymbol{E}) = \tfrac{1}{4}(\boldsymbol{\Sigma} \times \boldsymbol{A}^{-1} + \boldsymbol{\Sigma} * \boldsymbol{A}^{-1} + \boldsymbol{A}^{-1} * \boldsymbol{\Sigma} + \boldsymbol{A}^{-1} \times \boldsymbol{\Sigma}).$$

That $\hat{\boldsymbol{E}}$ is normally distributed with the indicated mean and covariance matrices can be shown directly, as in Ch. 2 of Anderson (1958). From (7) it is obvious that the linear elastic, isotropic stress matrix $\boldsymbol{T} = \lambda(\operatorname{tr} \boldsymbol{E})\boldsymbol{I} + 2\mu\boldsymbol{E}$ is normally distributed, since each component of the stress matrix is a linear combination of normally distributed variables, when the material constants λ and μ are known. We forbear writing out the details.

5. AN ESTIMATOR FOR THE CAUCHY-GREEN TENSOR $\boldsymbol{B}$

The estimator $\hat{\boldsymbol{F}}$ is distributed as a multivariate normal variable $N(\boldsymbol{F}, \boldsymbol{\Sigma} \times \boldsymbol{A}^{-1})$. We now seek an estimator for the matrix $\boldsymbol{B} = \boldsymbol{F}\boldsymbol{F}^T$. From (10) we see that

$$\mathscr{E}(\hat{\boldsymbol{F}} - \boldsymbol{F})(\hat{\boldsymbol{F}} - \boldsymbol{F})^T = \operatorname{tr}(\boldsymbol{A}^{-1}\boldsymbol{\Sigma}),$$

where tr is the matrix trace operator. Hence we see that

$$\begin{aligned}\mathscr{E}(\hat{\boldsymbol{F}}\hat{\boldsymbol{F}}^T) &= \mathscr{E}\{(\hat{\boldsymbol{F}} - \boldsymbol{F})(\hat{\boldsymbol{F}} - \boldsymbol{F})^T + \boldsymbol{F}\hat{\boldsymbol{F}}^T + \hat{\boldsymbol{F}}\boldsymbol{F}^T - \boldsymbol{F}\boldsymbol{F}^T\} \\ &= \operatorname{tr}(\boldsymbol{A}^{-1})\boldsymbol{\Sigma} + \boldsymbol{B}.\end{aligned}$$

Since $n\hat{\boldsymbol{\Sigma}}$ has the Wishart distribution $W(\boldsymbol{\Sigma}, n-3)$,

$$\mathscr{E}(n\hat{\boldsymbol{\Sigma}}) = (n-3)\boldsymbol{\Sigma},$$

and so

$$\hat{\boldsymbol{B}} \equiv \hat{\boldsymbol{F}}\hat{\boldsymbol{F}}^T - \frac{n}{n-3}\operatorname{tr}(\boldsymbol{A}^{-1})\hat{\boldsymbol{\Sigma}} \tag{11}$$

is an unbiased estimator for $\boldsymbol{B}$. Let us now examine the distribution of this statistic.

First let us rewrite (11) as

$$\hat{\boldsymbol{B}} = \hat{\boldsymbol{F}}\hat{\boldsymbol{F}}^T - \frac{n}{n-3}\operatorname{tr}(\boldsymbol{A}^{-1})\hat{\boldsymbol{\Sigma}}$$

$$= Y(X^TA^{-2}X)Y^T - \frac{1}{n-3}\text{tr}(A^{-1})(YI_nY^T - YX^T A^{-1}XY^T),$$

where we have used (8) and (9), and I_n is the $n \times n$ identity matrix; and we go on to write

$$\hat{B} = YMY^T,$$

where

$$M \equiv X^TA^{-2}X - \frac{1}{n-3}\text{tr}(A^{-1})(I_n - X^TA^{-1}X).$$

The matrix M is a mathematical matrix, in the sense that all its components are functions of the mathematical variables $u_{i\alpha}$, and so it can be regarded as known; it is also symmetric, and square of order n.

The matrix Y has covariance matrix $\Sigma \times I_n$ of order $3n$, and each element $\hat{B}_{ij}$ of $\hat{B}$ is a quadratic form in the normal random variables v_{ij} of Y; in fact,

$$\hat{B}_{ij} = v_{i\alpha}M_{\alpha\beta}v_{j\beta}.$$

Define the square symmetric matrix S_{ij} of order $3n$ by

$$S_{ij} \equiv \tfrac{1}{2}(E_{ij} + E_{ji}) \times M$$

where E_{ij} is the 3×3 matrix with zeros everywhere except at the intersection of the ith row and the jth column, which component is unity. Write Y as the (row) vector

$$V^T = [v_{11}, v_{21}, v_{31}, v_{12}, v_{22}, v_{32}, \cdots v_{1n}, v_{2n}, v_{3n}],$$

and then it is easy to see that

$$\hat{B}_{ij} = V^TS_{ij}V.$$

Theorem 1 of Baldessari (1967) now applies directly. Each $\hat{B}_{ij}$ is distributed as the weighted sum of noncentral chi-square (χ'^2) variates with various degrees of freedom and noncentrality parameters. In detail, each $\hat{B}_{ij}$ is a sum

$$\sum_{k=1}^{s_{ij}} a_{ijk}\, \chi'^2(r_{ijk}, \tfrac{1}{2} f^T L_{ijk} f)$$

where $a_{ijk} \neq a_{ijm}$ for $k \neq m$, and the a_{ijk} are the distinct eigenvalues of S_{ij} $(\Sigma \times I_n)$, with multiplicities r_{ijk}; the matrix L_{ijk} is positive semidefinite, symmetric, of order 3n and rank r_{ijk}, $k = 1, 2, \cdots, s_{ij}$, and this matrix can be taken as

$$L_{ijk} = E_{ijk}(\Sigma^{-1} \times I_n)$$

where S_{ij} has the spectral decomposition

$$S_{ij} = \sum_{k=1}^{s_{ij}} a_{ijk}E_{ijk},$$

with rank $(E_{ijk}) = r_{ijk}$, $k = 1, 2, \cdots, s_{ij}$, where the E_{ijk} satisfy

$$E_{ijk}E_{ijm} = E_{ijk}\delta_{km} \text{ (no sum on } i,j).$$

The vector f is the (row) vector corresponding to the matrix FX, i.e.,

$$f^T = [F_{1j}u_{j1}, F_{2j}u_{j1}, F_{3j}u_{j1}, F_{1j}u_{j2}, \cdots, F_{3j}u_{jn}].$$

REFERENCES

ANDERSON, T. W. (1958) *An Introduction to Multivariate Statistical Analysis.* New York: Wiley.

BALDESSARI, B. (1967). The distribution of a quadratic form of normal random variables. *Ann. Math. Statist.* **38,** 1700–4.

DALLY, J. W. & RILEY, W. F. (1965). *Experimental Stress Analysis.* New York: McGraw-Hill.

HICKSON, V. M. (1965). Some new techniques in strain measurement — a survey. In *Stress Analysis,* eds. O. C. Zienkiewicz and G. S. Hollister, 239–63 London: Wiley.

PERRY, C. C. & LISSNER, H. R. (1962). *The Strain Gage Primer.* New York: McGraw-Hill.

SANCHEZ, J. C. & WRIGHT, W. V. (1962). *Semiconductor and Conventional Strain Gauges.* London: Academic Press.

SOKOLNIKOFF, I. S. (1956). *Mathematical Theory of Elasticity.* New York: McGraw-Hill.

TRUESDELL, C. & NOLL, W. (1965). *The Non-linear Field Theories of Mechanics.* Handbuch der Physik, Band III/3. Berlin: Springer-Verlag.

TURKIN, B. V. (1972). Izmereniya polozhenii geometricheskikh figur v prostranstve kak odna iz zadach izmeritel'nykh laboratorii. *Izmeritel. Tekh.* **6,** 38–40.

The upcrossing rate of a class of stochastic processes

A. M. HASOFER
University of New South Wales

1. Introduction

The rate of upcrossings at level x, q_x, of a stationary stochastic process $X(t)$ is given, under fairly weak conditions, by the formula

$$q_x = \int_0^\infty y p(x, y)\, dy, \tag{1}$$

where $p(x, y)$ is the joint density function of $X(t)$ and its derivative $X'(t)$.

Sufficient conditions are:

(i) $X(t)$ has almost surely a continuous derivative $X'(t)$;

(ii) let $p_\tau(x, y)$ be the joint density function of $X(t)$ and $[X(t+\tau) - X(t)]/\tau$. Then

$$\lim_{\tau \to 0} \int_0^\infty dy \int_0^y p(x + \tau z, y) dz = \int_0^\infty y p(x, y) dy .$$

In particular, this will be true if

$$\text{(a)} \quad \lim_{\tau \to 0} p_\tau(x + \tau z, y) = p(x, y)$$

for all fixed x, y, z;

$$\text{(b)} \quad p_\tau(x, y) \leqq \Phi(y) \text{ for all } x \text{ and all } y > 0,$$

and $\int_0^\infty y\Phi(y) dy < \infty$ (by dominated convergence).

The above results follow immediately from Cramér & Leadbetter (1967, Chapter 10).

In this paper we focus our attention on the calculation of q_x by means of formula (1) for a comparatively large class of stochastic processes which are of interest in electrical and civil engineering. The implications of the sufficient conditions given will be dealt with elsewhere.

The only processes, as far as the author knows, for which q_x has been calculated so far are

(a) Gaussian processes,

(b) the envelope of a Gaussian process, and

(c) the "Rayleigh Process"

$$X(t) = [U_1^2(t) + U_2^2(t)]^{\frac{1}{2}},$$

where $U_1(t)$ and $U_2(t)$ are independent stationary Gaussian processes with mean zero and identical covariance functions (Hasofer, 1970).

Both (a) and (b) are quoted in Cramér & Leadbetter (1967). A simpler derivation of (b) is given Hasofer (1973).

The technique described in this paper is a generalization of the one used in Hasofer (1970) and Hasofer (1973).

2. THE MODEL

Let $\{U_i(t);\ i = 1, \cdots, n\}$ be a family of stochastic processes. In all but one of the cases considered, the $U_i(t)$ will be assumed to be Gaussian.

We are interested in calculating the rate of upcrossings of the derived stochastic process

$$X(t) = \phi[U_1(t), \cdots, U_n(t)], \tag{2}$$

where ϕ is a function of n variables having continuous partial derivatives.

We shall consider the following examples:

(i) $$X(t) = U_1(t) + U_2(t),$$

where $U_1(t)$, $U_2(t)$ are arbitrary, independent stochastic processes.

(ii) $$X(t) = U_1^2(t) + \cdots + U_n^2(t),$$

where the $U_i(t)$ are Gaussian processes.

(iii) $$X(t) = U_1(t)\,|U_1(t)| + \cdots + U_n(t)\,|U_n(t)|.$$

Example (ii) can be thought of as an approximation for the output of a "square law detector" followed by a linear filter, which is given by

$$X(t) = \int_{-\infty}^{+\infty} h(t-u)U^2(u)du,$$

where $U(t)$ is the input, and $h(t)$ the impulse response function of the filter, while example (iii) is an approximation to the stress on a structural member of a building due to a Gaussian wind $U(t)$, given by

$$X(t) = \int_{-\infty}^{+\infty} h(t-u)U(u)\,|U(u)|\,du$$

(see Hasofer, 1972).

3. GENERAL THEORY

The process $X(t)$ has an a.s. derivative $Y(t)$ given by

$$Y(t) = X_1U_1'(t) + \cdots + X_nU_n'(t), \tag{3}$$

where

$$X_i = \frac{\partial X}{\partial U_i}, \quad i = 1, \cdots, n.$$

Let the joint density function of X and Y be $p(x, y)$, and let

$$g(x) = \int_{-\infty}^{+\infty} p(x, y)dy$$

be the marginal density function of $X(t)$. Then $p(x, y) = g(x)p(y\,|\,x)$, where $p(y\,|\,x)$ is the density function of Y given that $X = x$.

We rewrite formula (1) as follows:

$$\begin{aligned} q_x &= g(x) \int_0^\infty y p(y \mid x)\, dy \\ &= g(x) E[Y_+ \mid X = x], \end{aligned}$$

where $E[Z \mid X = x]$ means the value of $E[Z \mid X]$ when $X = x$, and $Y_+ = \max(0, Y)$.

Now

$$E[Y_+ \mid X = x] = E\{E[Y_+ \mid U_1, \cdots, U_n] \mid X = x\}.$$

Suppose

$$E[Y_+ \mid U_1, \cdots, U_n] = h(U_1, \cdots, U_n).$$

We make the transformation of variables

$$\begin{aligned} X &= \phi(U_1, \cdots, U_n), \\ V_2 &= \phi_2(U_1, \cdots, U_n), \\ &\cdots\cdots\cdots\cdots \\ V_n &= \phi_n(U_1, \cdots, U_n). \end{aligned}$$

The functions $\phi_2, \cdots, \phi_n$ are chosen so that the transformation has a unique inverse

$$\begin{aligned} U_1 &= \psi_1(X, V_2, \cdots, V_n), \\ &\cdots\cdots\cdots\cdots \\ U_n &= \psi_n(X, V_2, \cdots, V_n). \end{aligned} \tag{4}$$

Let the joint density function of $U_1, \cdots, U_n$ be $p(u_1, \cdots, u_n)$, and the joint density function of $X, V_2, \cdots, V_n$ be $q(x, v_2, \cdots, v_n)$. Then

$$q(x, v_2, \cdots, v_n) = p(u_1, \cdots, u_n) \mid J_n \mid$$

where J_n is the jacobian of the transformation (4). The density function of $V_2, \cdots, V_n$ given that $X = x$ is given by $q(x, v_2, \cdots, v_n)[g(x)]^{-1}$.

Finally

$$\begin{aligned} q_x &= g(x) \int_{-\infty}^{+\infty} \cdots \int_{-\infty}^{+\infty} h(u_1, \cdots, u_n)\, q(x, v_2, \cdots, v_n)[g(x)]^{-1} dv_2 \cdots dv_n \\ &= \int_{-\infty}^{+\infty} \cdots \int_{-\infty}^{+\infty} h(u_1, \cdots, u_n)\, q(x, v_2, \cdots, v_n) dv_2 \cdots dv_n \end{aligned}$$

where $u_i = \psi_i(x, v_2, \cdots, v_n) \qquad (i = 1, \cdots, n)$.

Suppose now that the $U_i(t)$ are stationary Gaussian processes with mean μ_i. Let $\boldsymbol{U} = (U_1, \cdots, U_n)^T$ and $\boldsymbol{U}' = (U_1', \cdots, U_n')^T$. Let the covariance function Σ of $(\boldsymbol{U}^T, \boldsymbol{U}'^T)^T$, be partitioned as

$$\Sigma = \begin{pmatrix} \Sigma_{11} & \Sigma_{12} \\ \Sigma_{21} & \Sigma_{22} \end{pmatrix}.$$

The joint density function of $\boldsymbol{U}'$, given $\boldsymbol{U}$, is multivariate normal with mean $\Sigma_{21}\Sigma_{11}^{-1}(\boldsymbol{U}-\boldsymbol{\mu})$, where $\boldsymbol{\mu}=(\mu_1,\cdots,\mu_n)^T$, and covariance $\Sigma_{22.1}=\Sigma_{22}-\Sigma_{21}\Sigma_{11}^{-1}\Sigma_{12}$ (see Anderson, 1958, p. 29). This follows from the fact that $E(\boldsymbol{U}')=0$.

We deduce that the density function of $Y(t)$, given $\boldsymbol{U}$, is normal with mean $\alpha=\boldsymbol{X}^T\Sigma_{21}\Sigma_{11}^{-1}(\boldsymbol{U}-\boldsymbol{\mu})$, variance $\sigma^2=\boldsymbol{X}^T\Sigma_{22.1}\boldsymbol{X}$, where $\boldsymbol{X}=(X_1,\cdots,X_n)^T$.

Then

$$E[Y_+\,|\,U_1,\cdots,U_n]=(2\pi)^{-\frac{1}{2}}\sigma^{-1}\int_0^\infty y\exp\left[-\tfrac{1}{2}\left(\frac{y-\alpha}{\sigma}\right)^2\right]dy$$

$$=(2\pi)^{-\frac{1}{2}}\sigma\exp\left(-\frac{1}{2}\frac{\alpha^2}{\sigma^2}\right)+\alpha\left[1-\Phi\left(-\frac{\alpha}{\sigma}\right)\right],$$

where $\Phi(x)$ is the distribution function of the standard normal variable.

Example 1. Sum of two independent processes. $U_1(t)$ and $U_2(t)$ are two independent stochastic processes. Let $X(t)=U_1(t)+U_2(t)$,

then

$$Y(t)=X'(t)=U_1'(t)+U_2'(t).$$

Let the joint density function of $U_1(t)$ and $U_1'(t)$ be $p_1(u_1,v_1)$ and that of $U_2(t)$, $U_2'(t)$ be $p_2(u_2,v_2)$, let the marginal density of $U_1(t)$ and $U_2(t)$ be $p_1(u_1)$, $p_2(u_2)$ respectively.

Then

$$E(Y_+\,|\,U_1,U_2)=[p_1(u_1)p_2(u_2)]^{-1}\int_0^\infty y\,dy\int_{-\infty}^{+\infty}p_1(u_1,y-z)p_2(u_2,z)dz\,.$$

We now make the transformation

$$X=U_1+U_2,$$

$$V_2=U_2.$$

Then

$$q(x,v_2)=p_1(x-v_2)p_2(v_2),$$

and

$$q_x=\int_0^\infty y\,dy\int_{-\infty}^{+\infty}\int_{-\infty}^{+\infty}p_1(x-v_2,y-z)p_2(v_2,z)dz\,dv_2\,.$$

In this case the formula could have been obtained by noting that

$$p(x,y)=\int_{-\infty}^{+\infty}\int_{-\infty}^{+\infty}p_1(x-v_2,y-z)p_2(v_2,z)dzdv_2\,,$$

since $U_1(t)$ and $U_2(t)$ are independent.

Example 2. $$X(t)=U_1^2(t)+\cdots+U_n^2(t),$$

where $U_1(t),\cdots,U_n(t)$ are jointly Gaussian.

Then

$$Y(t) = 2U_1(t)U_1'(t) + \cdots + 2U_n(t)U_n'(t).$$

We make the transformation

$$U_1 = X^{\frac{1}{2}} \cos\theta_1 \cdots \cos\theta_{n-1}$$

$$U_2 = X^{\frac{1}{2}} \cos\theta_1 \cdots \cos\theta_{n-2} \sin\theta_{n-1}$$

$$. \quad . \quad . \quad . \quad . \quad . \quad . \quad . \quad . \quad .$$

$$U_j = X^{\frac{1}{2}} \cos\theta_1 \cdots \cos\theta_{n-j} \sin\theta_{n-j+1}$$

$$. \quad . \quad . \quad . \quad . \quad . \quad . \quad .$$

$$U_n = X^{\frac{1}{2}} \sin\theta_1,$$

where the range of the θ_i is $(-\pi/2, \pi/2)$ for $i = 1, \cdots, n-2$ and $(0, 2\pi)$ for $i = n-1$.

The Jacobian is

$$J = \tfrac{1}{2}X^{(n-2)/2} \cos^{n-2}\theta_1 \cos^{n-3}\theta_2 \cdots \cos\theta_{n-2}$$

$$= X^{(n-2)/2} B(\theta_1, \cdots, \theta_n), \text{ say.}$$

The joint density function of $U_1, \cdots, U_n$ is

$$(2\pi)^{-n/2} |\Sigma_{11}|^{-\frac{1}{2}} \exp[-\tfrac{1}{2}(\boldsymbol{u} - \boldsymbol{\mu})'\Sigma_{11}^{-1}(\boldsymbol{u} - \boldsymbol{\mu})]$$

which is of the form

$$(2\pi)^{-n/2} |\Sigma_{11}|^{-\frac{1}{2}} \exp[-\tfrac{1}{2}A(x, \theta_1, \cdots, \theta_n)].$$

Thus

$$q_x = (2\pi)^{-(n+1)/2} |\Sigma_{11}|^{-\frac{1}{2}} \int_{-\frac{1}{2}\pi}^{\frac{1}{2}\pi} \cdots \int_{-\frac{1}{2}\pi}^{\frac{1}{2}\pi} \int_0^{2\pi} \left\{\sigma \exp\left(-\frac{1}{2}\frac{\alpha^2}{\sigma^2}\right) + (2\pi)^{\frac{1}{2}}\alpha\left[1 - \Phi\left(-\frac{\alpha}{\sigma}\right)\right]\right\}$$

$$\exp[-\tfrac{1}{2}A(x, \theta_1, \cdots, \theta_n)] x^{(n-2)/2} B(\theta_1, \cdots, \theta_n) d\theta_1 \cdots d\theta_{n-1},$$

where

$$\alpha = 2\boldsymbol{u}' \Sigma_{21}\Sigma_{11}^{-1}(\boldsymbol{u} - \boldsymbol{\mu}),$$

$$\sigma^2 = 4\boldsymbol{u}'(\Sigma_{22} - \Sigma_{21}\Sigma_{11}^{-1}\Sigma_{12})\boldsymbol{u}.$$

The formula simplifies enormously if we assume that the $U_i(t)$ are independent, identically distributed Gaussian processes with mean zero and marginal variance unity. Then $\alpha = 0$, $\sigma^2 = 4\lambda_2(u_1^2 + \cdots + u_n^2) = 4\lambda_2 x$, where λ_2 is the marginal variance of the $U_i'(t)$.

It follows that

$$h(u_1, \cdots, u_n) = \left(\frac{4\lambda_2 x}{2\pi}\right)^{\frac{1}{2}},$$

which does not depend on $\theta_1, \cdots, \theta_n$. Thus

$$q_x = 2\left(\frac{\lambda_2 x}{2\pi}\right)^{\frac{1}{2}} g(x),$$

where $g(x)$ is the density function of $X(t)$;

$$g(x) = \frac{1}{2^{\frac{1}{2}n}\Gamma(\frac{1}{2}n)}\exp(-\tfrac{1}{2}x)x^{\frac{1}{2}n-1}.$$

Thus

$$q_x = \left(\frac{\lambda_2}{2\pi}\right)^{\frac{1}{2}} \frac{1}{2^{\frac{1}{2}n-1}\Gamma(\frac{1}{2}n)} \exp(-\tfrac{1}{2}x)\, x^{(n-1)/2}.$$

As another special case of the same example, one can obtain the upcrossing rate of the square of the envelope of a stationary Gaussian process $U_1(t)$ with mean zero.

Let

$$X(t) = U_1^2(t) + U_2^2(t),$$

where $U_2(t)$ is the Hilbert transform of $U_1(t)$.

The matrix Σ in this case is of the form:

$$\begin{bmatrix} \lambda_0 & 0 & 0 & \lambda_1 \\ 0 & \lambda_0 & -\lambda_1 & 0 \\ 0 & -\lambda_1 & \lambda_2 & 0 \\ \lambda_1 & 0 & 0 & \lambda_2 \end{bmatrix}$$

while $\boldsymbol{\mu} = 0$ (see Hasofer, 1973).

It turns out that $\alpha = 0$ and $\sigma^2 = (\lambda_2 - \lambda_1\lambda_0^{-1})x$. It then follows easily that

$$q_x = \left(\frac{\lambda_2 - \lambda_1\lambda_0^{-1}}{2\pi}\right)^{\frac{1}{2}} \frac{x^{\frac{1}{2}}}{\lambda_0} \exp\left(-\frac{1}{2}\frac{x^{\frac{1}{2}}}{\lambda_0}\right).$$

Example 3.

Let $X(t) = U_1(t)\,|U_1(t)| + \cdots + U_n(t)\,|U_n(t)|$.

It is easy to verify that, if $y = x|x|$, then $dy/dx = 2|x|$. Therefore

$$Y(t) = 2|U_1(t)|\,U_1'(t) + \cdots + 2|U_n|(t)U_n'(t).$$

A suitable transformation of variables is

$$\begin{aligned} x &= u_1|u_1| + \cdots + u_n|u_n|, \\ v_2 &= u_2|u_2|, \\ &\cdot \quad \cdot \quad \cdot \quad \cdot \\ v_n &= u_n|u_n|. \end{aligned}$$

The inverse transformation is

$$\begin{aligned} u_1 &= (x - v_2 \cdots - v_n)\,|x - v_2 \cdots - v_n|^{-\frac{1}{2}}, \\ u_2 &= v_2|v_2|^{-\frac{1}{2}}, \\ &\cdot \quad \cdot \quad \cdot \quad \cdot \\ u_n &= v_n|v_n|^{-\frac{1}{2}}, \end{aligned}$$

and the Jacobian is

$$J(x, v_2, \cdots, v_n) = 2^{-n}|x - v_2 \cdots - v_n|^{-\frac{1}{2}}|v_2|^{-\frac{1}{2}} \cdots |v_n|^{-\frac{1}{2}}.$$

Finally

$$q_x = \int_{-\infty}^{+\infty} \cdots \int_{-\infty}^{+\infty} h(u_1, \cdots, u_n)\, p(u_1, \cdots, u_n)\, J(x, v_2, \cdots, v_n) dv_2 \cdots dv_n.$$

REFERENCES

ANDERSON, T. W. (1958). *An Introduction to Multivariate Statistical Analysis.* New York: Wiley.

CRAMÉR, H. & LEADBETTER, M. R. (1967). *Stationary and Related Stochastic Processes.* New York: Wiley.

HASOFER, A. M. (1970). On the derivative and the upcrossings of the Rayleigh process. *Aust. J. Statist.* **12,** 150–1.

HASOFER, A. M. (1972). Wind-load design based on statistics. *Proc. Inst. Civ. Eng.* **51,** 69–82.

HASOFER, A. M. (1973). The joint distribution of the envelope of a Gaussian process and its derivative. *Aust. J. Statist.* **15.**

Mathematical models in ecology*

D. R. McNEIL
Princeton University

1. Introduction and summary

Ecology is concerned with the interaction between living organisms and their environment. Mathematical ecologists are interested in theories, or mathematical models, to account for observed ecological phenomena, such as the relative abundances of different species of animals and plants, their spatial distributions and temporal fluctuations, including periodic cycles.

In this paper we first survey some of the mathematical models which have been proposed to describe the growth and interaction of species. We then show how they may be extended to allow for stochastic variation. The stochastic models are of two kinds: (a) those which take account of the finiteness and resulting extinction probability of the population, and (b) those which allow for random changes in the environment. The results may be used to determine what is gained by introducing randomness into the model.

2. Deterministic models for isolated populations

The simplest mathematical model describing population growth is the geometric growth or Malthusian model

$$y_{t+1} = ay_t, \tag{2.1}$$

where y_t is the population at time t and a is the growth ratio from one time point to the next, assumed constant. Equation (2.1) is a discrete generation model, in contrast to the analogous continuous generation model

$$\dot{y}(t) = \alpha y(t) \quad . \qquad (\dot{y} = dy/dt) \tag{2.2}$$

The solutions to (2.1) and (2.2), however, are essentially equivalent, being $y_t = y_0 a^t$ and $y(t) = y(0)e^{\alpha t}$, respectively. Thus they are identical if $\alpha = \log a$. It may be noted that the population growth is positive or negative according as $a > 1$ or $a < 1$ (or $\alpha > 0$ or $\alpha < 0$), respectively.

Equations (2.1) and (2.2) describe uninhibited growth (or decay, if $a < 1$); in practice it is more realistic to assume that the growth rate is density-dependent or dependent on the population size at time t. Thus we should replace equations (2.1) and (2.2) by

*Research supported by grants from the Ford and Rockefeller Foundations, awarded the Office of Population Research, Princeton University.

$$y_{t+1} = a(y_t)y_t, \tag{2.3}$$

and

$$\dot{y}(t) = \alpha\{y(t)\}y(t), \tag{2.4}$$

respectively. Again the choice of (2.3) or (2.4) depends partly on whether we are interested in discrete or continuous generation, but an extra factor emerges—mathematical ease of solution. In general, equation (2.3) does not yield a simple explicit solution, although it is easy to compute y_t by iteration. In contrast, the solution to (2.4) is not directly computable, but an approximate formula may be obtained by assuming that the density-dependence in the growth rate is slight. Writing $\alpha(y) = \alpha - \beta y$, equation (2.4) yields

$$y(t) = \frac{\alpha}{\beta}\left\{1 + \left(\frac{\alpha}{\beta y(0)} - 1\right)e^{-\alpha t}\right\}^{-1}. \tag{2.5}$$

Evidently $y(t) \rightarrow \alpha/\beta$ as $t \rightarrow \infty$, the convergence being exponential.

The model with $\alpha(y) = \alpha - \beta y$ is known as the *logistic* model for population growth, and is important in ecology for several reasons. First, as stated above, the linear form of the growth rate may be regarded as a first approximation to the true situation. Second, the parameters α and β, and the ratio α/β, all have simple ecological interpretations; α is the growth rate when the population is small or unrestricted, β is the crowding effect, or more precisely the amount by which the growth rate is further reduced by the addition of a single individual, and α/β is the limiting population size, or the capacity of the environment being inhabited. Last but not least the linear form of $\alpha(y)$ yields the simple closed form solution (2.5).

For populations described by equations (2.3) or (2.4), the first question of interest is that of stability. If the population eventually settles down to a constant size, $\bar{y}$, then $\bar{y}$ will satisfy $a(\bar{y}) = 1$ or $\alpha(\bar{y}) = 0$. We may examine the stability of the solution $\bar{y}$ as follows. Taking the discrete generation model, we perturb y_t about $\bar{y}$, so that $y_t = \bar{y} + \varepsilon u_t$. Then (2.3) gives, using Taylor's expansion

$$\begin{aligned} \bar{y} + \varepsilon u_{t+1} &= a(\bar{y} + \varepsilon u_t)(\bar{y} + \varepsilon u_t) \\ &= a(\bar{y})\bar{y} + \varepsilon u_t\{a(\bar{y}) + a'(\bar{y})\bar{y}\} + o(\varepsilon), \end{aligned}$$

whence

$$u_{t+1} = \{1 + a'(\bar{y})\bar{y}\}u_t, \tag{2.6}$$

on setting $a(\bar{y}) = 1$, dividing by ε, and taking the limit as $\varepsilon \rightarrow 0$. Comparing equations (2.6) and (2.2), it follows that small disturbances from $\bar{y}$ will die out provided that $-2 < \bar{y}a'(\bar{y}) < 0$. This condition, together with $\bar{y} > 0$, is necessary for $\bar{y}$ to be a stable equilibrium. The analogous condition in the continuous generation case is $\bar{y} > 0$ and $\alpha'(\bar{y}) < 0$.

3. Deterministic models for species interaction

The extension of equation (2.1) to the case of a number of interacting species is

$$\boldsymbol{y}_{t+1} = \boldsymbol{A}\boldsymbol{y}_t, \tag{3.1}$$

where $\boldsymbol{y}_t$ is the vector whose components are the species abundances at time t, and $\boldsymbol{A}$ is a square matrix. Equation (3.1) is the basic model of human population projection by age groups; in this case the components of $\boldsymbol{y}_t$ are the numbers in the various age-groups. $\boldsymbol{A}$ is called the projection matrix or Leslie matrix after Leslie (1945). However, equation (3.1) is really a discrete approximation to a model originally studied by Sharpe and Lotka (1911).

The asymptotic properties of the age distribution $\boldsymbol{y}_t$ are determined by the largest (in absolute value) eigenvalue of $\boldsymbol{A}$. In demographic applications it may be shown that this eigenvalue, λ, say, is unique and positive, in which case equation (3.1) yields

$$\boldsymbol{y}_t = \lambda^t\{\boldsymbol{\eta} + o(1)\}, \tag{3.2}$$

where $\boldsymbol{\eta}$ is the stable age distribution. Thus, as in the univariate case, the long-term growth is geometric, and positive or negative according as $\lambda > 1$ or $\lambda < 1$. In general, the solution to (3.1) tends to the null vector $\boldsymbol{0}$ as $t \to \infty$ if, and only if, all the eigenvalues of $\boldsymbol{A}$ are less than unity in absolute value.

While equation (3.1) provides an important model for demographic projection and analysis, its application to ecology is limited, as in the univariate case, by the constant growth-rate assumption. Accordingly, we progress to the general nonlinear model

$$\boldsymbol{y}_{t+1} = \boldsymbol{A}(\boldsymbol{y}_t)\boldsymbol{y}_t. \tag{3.3}$$

The search for equilibrium solutions proceeds exactly as in the univariate case. For a steady state we need to find $\bar{\boldsymbol{y}}$, with $\bar{y}_i \geqq 0$ for all i and $\bar{y}_i > 0$ for at least one i, satisfying $\boldsymbol{A}(\bar{\boldsymbol{y}})\bar{\boldsymbol{y}} = \bar{\boldsymbol{y}}$. To examine the stability, we put $\boldsymbol{y}_t = \bar{\boldsymbol{y}} + \varepsilon \boldsymbol{u}_t$ in equation (3.3), use Taylor's expansion, and obtain, as in the univariate case,

$$\boldsymbol{u}_{t+1} = \boldsymbol{F}(\bar{\boldsymbol{y}})\boldsymbol{u}_t, \tag{3.4}$$

where $\boldsymbol{F}(\bar{\boldsymbol{y}}) = \{\partial/\partial\bar{\boldsymbol{y}}\}\{\boldsymbol{A}(\bar{\boldsymbol{y}})\bar{\boldsymbol{y}}\}$.

By comparison with equation (3.1), equation (3.3) has a stable solution $\bar{\boldsymbol{y}}$ if $\bar{\boldsymbol{y}}$ has at least one positive component and no negative components, and $\boldsymbol{F}(\boldsymbol{y})$ has all its eigenvalues within the unit circle. Any components of $\bar{\boldsymbol{y}}$ which are zero correspond to species which fail to survive.

The continuous generation vector model

$$\dot{\boldsymbol{y}}(t) = \boldsymbol{A}(\boldsymbol{y}(t))\boldsymbol{y}(t) \tag{3.5}$$

may be analysed by the same methods. In this case it turns out that the condition for a stable equilibrium is that the eigenvalues of $\{\partial/\partial\bar{\boldsymbol{y}}\}\{\boldsymbol{A}(\bar{\boldsymbol{y}})\bar{\boldsymbol{y}}\}$ be in the left half-plane.

4. Volterra models

Dynamic models for the relationship between two species were first considered by Lotka (1925) and independently by Volterra (1931), who proposed the continuous generation equations

$$\begin{aligned}\dot{y}_1(t) &= y_1(t)\{\alpha_1 - \beta_1 y_1(t) - \gamma_1 y_2(t)\},\\ \dot{y}_2(t) &= y_2(t)\{\alpha_2 - \beta_2 y_2(t) - \gamma_2 y_1(t)\}.\end{aligned} \tag{4.1}$$

Equations (4.1) constitute a natural generalization of the univariate logistic model (2.5). Here γ_1 and γ_2 are additional parameters representing the degree of overlap or interaction between the two species. Specifically, γ_1 is the amount by which an additional individual from the second species retards the growth rate of the first species. Of course γ_1 or γ_2 or both could be negative. This would occur in the case of cooperation or predation. It should be noted that equations (4.1) are less general than those obtained by taking the first approximation to (3.5) by including only the linear terms in $\boldsymbol{A}(\boldsymbol{y}(t))$.

While explicit solutions to equations (4.1) are not always possible, the limiting behaviour as $t \to \infty$ is interesting. It is easy to show that in the absence of crowding ($\beta_1 = \beta_2 = 0$) no stable equilibrium can exist, although a limit cycle results in this case if in addition $\gamma_1\gamma_2$, $\alpha_1\alpha_2$ and $\alpha_1\gamma_1$ are all negative as in a predator-prey situation. We may conclude that two species whose abundances satisfy equation (4.1) cannot survive in the same environment solely by each limiting the numbers of the other — each must be able to limit its own growth independently of the other.

This result is true for more general models, and for more than two interacting species, and is known as the *principle of competitive exclusion*. At first sight it appears confusing, since it seems to say that diversity is contrary to stability. The explanation lies in the fact that environments are heterogeneous, so that one species will tend to dominate in one section, another in a different section.

A natural generalization of equations (4.1) to more than two species is

$$\frac{d}{dt}\log \mathbf{y}(t) = \boldsymbol{\alpha} - \boldsymbol{B}\mathbf{y}, \tag{4.2}$$

where $\boldsymbol{\alpha}$ is the vector of unrestricted growth rates and $\boldsymbol{B}$ is the matrix whose (i,j) element is the amount by which an additional individual from species j reduces the growth rate of species i.

5. Species competing for resources

To generalize the Volterra models to allow for an inhomogeneous environment, MacArthur (1972) postulated a mathematical theory in which species compete for resources or foods which are spread non-uniformly over an environment. We denote by $r(\boldsymbol{x})$ the resource abundance at coordinate $\boldsymbol{x}$, where $\boldsymbol{x}$ has sufficiently many components to completely specify the position and type of resource. It is assumed that, in the absence of competing species, the resources, independently of each other, grow according to the logistic model (2.5), that is,

$$\dot{r}(t, \boldsymbol{x}) = a(\boldsymbol{x})r(t, \boldsymbol{x}) - b(\boldsymbol{x})r^2(t, \boldsymbol{x}). \tag{5.1}$$

To describe the effect of the species (or predators, from the point of view of the resources) we introduce a resource utilization vector, $\boldsymbol{c}(\boldsymbol{x})$, whose ith component specifies the contribution to the diet of species i by the resource at $\boldsymbol{x}$. The dynamics of growth of species i are now given by

$$\dot{y}_i(t) = \left\{ -\alpha_i + \delta_i \int c_i(\boldsymbol{x})r(t, \boldsymbol{x})d\boldsymbol{x} \right\} y_i(t), \tag{5.2}$$

where α_i is the rate at which the abundance of species i declines in the absence of resources, and δ_i is the rate at which species i is able to convert resource quantity into new individuals of its own species. The integration is over all resources in the environment.

Allowing for the depletion by the competitors, equation (5.1) should be replaced by

$$\dot{r}(t, \boldsymbol{x}) = \{a(\boldsymbol{x}) - \boldsymbol{c}^T(\boldsymbol{x})\boldsymbol{y}(t)\}r(t, \boldsymbol{x}) - b(\boldsymbol{x})r^2(t, \boldsymbol{x}), \tag{5.3}$$

where $\boldsymbol{c}^T$ denotes the transpose of $\boldsymbol{c}$. Equations (5.2), for $i = 1, 2, \cdots, k$, possess the matrix representation

$$\frac{d}{dt}\{\log \boldsymbol{y}(t)\} = -\boldsymbol{\alpha} + \boldsymbol{D}\int \boldsymbol{c}(\boldsymbol{x})r(t, \boldsymbol{x})d\boldsymbol{x}, \tag{5.4}$$

where $\boldsymbol{D}$ is the diagonal matrix with elements δ_i. Equating the left-hand sides of equations (5.3) and (5.4) to zero, we find that the equilibrium species abundances $\bar{\boldsymbol{y}}$ satisfy the linear equations

$$\int \boldsymbol{c}(x)\{a(x) - \boldsymbol{c}^T(x)\bar{\boldsymbol{y}}\}dx\,/b(x) = \boldsymbol{D}^{-1}\boldsymbol{\alpha}. \tag{5.5}$$

A very simple interpretation is possible when $\boldsymbol{D}^{-1}\boldsymbol{\alpha} = k\boldsymbol{e}$ where $\boldsymbol{e}$ is theunit vector and k is constant. (This is the case when each competitor has the same ratio of resource conversion rate to natural growth rate.) In this case it is easy to see that $\bar{\boldsymbol{y}}$ is the value of $\boldsymbol{y}$ which minimizes the quadratic form

$$Q(\bar{\boldsymbol{y}}) = \tfrac{1}{2}\int \{a(x) - kb(x) - \boldsymbol{c}^T(x)\boldsymbol{y}\}^2 dx\,/b(x). \tag{5.6}$$

The interpretation of Q is as a weighted integral, or average, of the squared differences between the resource growth rates (allowing for the effect of crowding) and the total resource utilizations by all species at the various positions. It follows that Q must decrease as time progresses, and that a new species can invade only if the resulting minimized value of Q is smaller than the previous minimized ralue. *A fortiori* a community in which $Q = 0$ cannot be invaded by *any* species.

6. Simple finite population models

Stochastic models for population growth date back to Galton & Watson (1874), who considered the problem of extinction of family names and thus initiated the theory of branching processes. Subsequently, McKendrick (1914) investigated continuous generation models for population growth. To describe these models, we start with the case of a univariate population for which the appropriate deterministic model would be either (2.1) or (2.2).

Considering (2.1) first, we define a sequence of random variables $\{Y_0, Y_1, Y_2, \cdots\}$ as follows. The conditional distribution of Y_{t+1}, given $Y_0, Y_1, Y_2, \cdots, Y_t$, depends only on Y_t, and has expectation aY_t and variance $\sigma^2 Y_t$. (Thus $\{Y_t\}$ is a Markov chain.) The following example will give some intuitive feeling for this model. We imagine that individual i in the population at generation t becomes ξ_i individuals at generation $t+1$, where $\{\xi_i, i = 1, 2, \cdots, Y_t\}$ is a set of independent, identically distributed random variables having expectations a and variances σ^2. The replacement of an individual by ξ individuals can occur in a variety of ways, for example, by each individual dying and being replaced by seeds as in the case of annual plants, or by occasional fission as with cell division or radioactive bombardment. It follows that, given Y_t, Y_{t+1} has expectation aY_t and variance $\sigma^2 Y_t$.

The precise distribution of Y_t may be complicated, but we can get simple asymptotic results, valid for large populations, as follows. Let the initial population size be Ny_0, where N is fixed and large and y_0 is also fixed. The expectation and variance of Y_t may be calculated by iteration; we obtain

$$E[Y_t] = aE[Y_{t-1}] = \cdots = a^t Y_0 = Ny_0 a^t, \tag{6.1}$$

$$\begin{aligned} \operatorname{var}[Y_t] &= \sigma^2 a^{t-1} Y_0 + a^2 \operatorname{var}[Y_{t-1}] \\ &= Ny_0\sigma^2 a^{t-1}(a^t - 1)/(a-1). \end{aligned} \tag{6.2}$$

Thus $Y_t/N \to y_t = y_0 a^t$ in probability as $N \to \infty$. Moreover, it may be shown that

$$Z_t = N^{\frac{1}{2}}\left\{\frac{1}{N}Y_t - y_t\right\} \tag{6.3}$$

has a limiting distribution as $N \to \infty$ which is Gaussian with mean zero and variance $y_0\sigma^2 a^{t-1}(a^t - 1)/(a-1)$. Furthermore, the sequence $\{Z_0 = 0, Z_1, Z_2, \cdots\}$ is a simple Markov chain in which

$$Z_{t+1} = aZ_t + \eta_t, \tag{6.4}$$

where η_t is Gaussian white noise with variance $\sigma^2 y_0$.

We conclude that, if N is sufficiently large, the population size at generation t is the superposition of (a) a deterministic part Ny_t, where y_t satisfies (2.1), and (b) a stochastic component $N^{\frac{1}{2}}Z_t$, where Z_t is the random walk satisfying (6.4).

Let us now investigate an analogous stochastic generalization of equation (2.2). We first define a birth and death process, $Y(t)$, that is, an integer valued, continuous-time Markov process with transition probabilities

$$\operatorname{pr}\{Y(t+h) = j \mid Y(t) = i\} = a_{ij}(h) + o(h), \tag{6.5}$$

where $a_{ij}(h) = \lambda_i h$ if $j = i+1$, $a_{ij}(h) = \mu_i h$ if $j = i-1$, $a_{ij}(h) = 1 - (\lambda_i + \mu_i)h$ if $j = i$, and $a_{ij}(h) = 0$ otherwise.

To obtain a stochastic version of the multiplicative model (2.2), we specialize (6.5) to $\lambda_i = \frac{1}{2}i(\beta + \alpha)$, $\mu_i = \frac{1}{2}i(\beta - \alpha)$. Intuitively, this would arise if, in each interval $(t, t+h)$, each individual in the population, independently of the others, has a chance $\frac{1}{2}(\beta + \alpha)h + o(h)$ of giving birth to a new individual and a chance $\frac{1}{2}(\beta - \alpha)h + o(h)$ of dying. (In general, λ_i/i and μ_i/i may be regarded as the birth and death rates per unit time *per individual*.) Then it follows that

$$\begin{aligned} E\{Y(t+h) - Y(t) \mid Y(t)\} &= \alpha Y(t)h + o(h), \\ \operatorname{var}\{Y(t+h) - Y(t) \mid Y(t)\} &= \beta Y(t)h + o(h). \end{aligned} \tag{6.6}$$

From equations (6.6) we deduce, after a few calculations,

$$\begin{aligned} E\{Y(t+h)\} - E\{Y(t)\} &= \alpha E\{Y(t)\}h + o(h), \\ \operatorname{var}\{Y(t+h)\} - \operatorname{var}\{Y(t)\} &= 2\alpha \operatorname{var}\{Y(t)\}h + \beta E\{Y(t)\}h + o(h). \end{aligned} \tag{6.7}$$

Dividing equation (6.7) by h, taking limits as $h \to 0$, and solving the resulting differential equations, we obtain

$$E\{Y(t)\} = Y(0)e^{\alpha t}, \quad \operatorname{var}\{Y(t)\} = \beta\alpha^{-1}Y(0)e^{\alpha t}(e^{\alpha t} - 1). \tag{6.8}$$

As for the discrete generation model, if $Y(0)$ is replaced by $Ny(0)$, it follows that $Y(t)/N$ converges in probability to $y(t) = y(0)\exp(\alpha t)$ as $N \to \infty$, and that

$$Z(t) = N^{\frac{1}{2}}\left\{\frac{1}{N}Y(t) - y(t)\right\} \tag{6.9}$$

is asymptotically Gaussian with mean zero and variance $y(0)\beta\alpha^{-1}e^{\alpha t}(e^{\alpha t} - 1)$. Using results in the theory of convergence of probability measures, which need not be discussed here (see Billingsley, 1968), it also may be shown that $Z(t)$ converges to a Brownian motion as $N \to \infty$.

This Brownian motion is a non-stationary Ornstein-Uhlenbeck diffusion process with instantaneous acceleration $\beta y(t)$ and drift $-\alpha Z(t)$.

Thus if N is large the population size at time t is the sum of the deterministic part $Ny(t)$, where $y(t)$ satisfies (2.2), and the stochastic part $N^{\frac{1}{2}}Z(t)$, where $Z(t)$ is the above Brownian motion.

We are now in a position to interpret the stochastic models which have been introduced. In both the discrete and continuous-generation models, the uncertainity is at the individual level. However, each individual is assumed to obey the same probability laws, and the uncertainty diminishes as the population size increases, and in the limit as it approaches infinity the stochastic and deterministic models become identical.

7. Simple random environment models

An alternative procedure for introducing random variation into a deterministic model is to assume that the parameters (for example, growth rates and crowding and interaction coefficients) are random variables. This approach was applied to the demographic model (3.1) by Sykes (1969). In the univariate case it involves replacing equation (2.1) by

$$Y_{t+1} = A_t Y_t, \tag{7.1}$$

where $\{A_0, A_1, A_2, \cdots\}$ is a sequence of random variables. If the A_t are identically distributed with mean a, then

$$E\{Y_{t+1}\} = aE\{Y_t\} + E\{(A_t - a)Y_t\}. \tag{7.2}$$

If, in addition, the A_t are uncorrelated, the second term on the right-hand side of equation (7.2) vanishes, and solving recursively we get (putting $Y_0 = y_0$, presumed fixed)

$$E\{Y_t\} = a^t y_0. \tag{7.3}$$

Similarly, the variance is given by

$$\operatorname{var}\{Y_t\} = \{(a^2 + \sigma^2)^t - a^{2t}\}y_0^2. \tag{7.4}$$

Thus the stochastic model obtained by adding random, uncorrelated fluctuations to the growth rate in equation (2.1) has the same mean as the deterministic model. As in the stochastic model for finite populations, much more general results are obtainable if the stochastic part is assumed relatively small. Accordingly, we suppose that $A_t = a + \varepsilon U_t$, where $\varepsilon > 0$ is small and U_t is a stationary time series.

Putting $A_t = a + \varepsilon U_t$ in equation (7.1), we get

$$\begin{aligned} Y_t &= (a + \varepsilon U_{t-1})(a + \varepsilon U_{t-2}) \cdots (a + \varepsilon U_0) y_0 \\ &= a^t y_0 + \varepsilon a^{t-1} y_0 \sum_{i=0}^{t-1} U_i + o(\varepsilon). \end{aligned} \tag{7.5}$$

Thus $Y_t \to y_t$ in probability as $\varepsilon \to 0$, where y_t is the deterministic model satisfying equation (2.1). Defining $Z_t = (Y_t - y_t)/\varepsilon$, it also follows provided the time series $\{U_t\}$ satisfies a central limit theorem, that Z_t has a limiting distribution as $\varepsilon \to 0$ which is approximately Gaussian with mean zero. The variance of this limiting distribution is $a^{2t-2}y_0^2 v_t$, where

$$v_t = \text{var}\left\{\sum_{i=0}^{t-1} U_i\right\} = \sum_{j=0}^{t-1} (t-j)\gamma_j . \qquad (\gamma_j = E\{U_t U_{t+j}\}) \tag{7.6}$$

A similar result holds for the continuous generation model, in which case we may write $Y(t) = y(t) + \varepsilon Z(t)$, where $Z(t)$ is convergent to a continuous-time non-stationary, approximately Gaussian process in the limit as $\varepsilon \to 0$.

We come then to an interpretation of the stochastic models for unrestricted population growth of isolated populations. There are two ways of introducing a stochastic component into the model. The first of these may be regarded as a finite population effect, while the second involves random fluctuations in the growth rate due to environmental effects.

In the first case the error is due to the unpredictability of the individual members of the population, but these errors tend to annihilate each other as the population size tends to infinity. Thus there is a *law of large numbers* convergence of the stochastic model to the deterministic model. Moreover, the population size at time t may be approximated by the superposition of the deterministic model and a relatively small Markovian Gaussian random process.

In the second type of stochastic model, the error is due to the unpredictability of the environment. Since these errors affect all the members of the population simultaneously, there is no tendency for them to cancel out. However, if the environmental fluctuation is small the error is additive, so that the population size at time t is the superposition of the deterministic model and a relatively small random process, as in the former case. This random process is, under mild restrictions, approximately Gaussian.

8. Stochastic logistic processes

The results of the last two sections carry over easily to more general forms of population growth such as those described by equations (2.3) and (2.4). Since there is no essential difference between discrete and continuous generation models, let us work henceforth in continuous time. To illustrate, we take the logistic process, which grows according to equation (2.5).

To obtain a finite population stochastic model, we need to define the sequences $\{\lambda_i\}$ and $\{\mu_i\}$ which arise in (6.5). Feller (1939) suggested that appropriate forms should be

$$\lambda_i = \alpha i(1 - i/N), \; \mu_i = (\beta - \alpha)i^2/N \qquad (\beta > \alpha). \tag{8.1}$$

In this formulation, it is clear that the population size can never exceed N, and the birth rate per individual decreases in proportion to the deficit below N while the death rate per individual increases in proportion to the population size. Although it is not possible to obtain simple explicit expressions for the mean and variance of the population size at time t, asymptotic results are available, which may be described as follows.

As in the case of the birth and death process corresponding to (2.2), it may be shown that $Y(t)/N \to y(t)$ as $N \to \infty$, where $y(t)$ is given by equation (2.5). Defining $Z(t)$ as in equation (6.9), it may be shown that as $N \to \infty$, $Z(t)$ converges in distribution to a Gaussian diffusion process with instantaneous acceleration $\frac{1}{2}\{\alpha y(t) - (2\alpha - \beta)y^2(t)\}$ and drift $-\{2\beta y(t) - \alpha\}Z(t)$. (For details of the derivation of these results, we refer to McNeil & Schach, 1973.)

It may be noted that, when the process is near its equilibrium, given by $y(t) = \alpha/\beta$ or $Y(t) = N\alpha/\beta$, the diffusion process characterizing the stochastic component of the population growth has instantaneous acceleration close to $\alpha^2(\beta - \alpha)/\beta^2$ and instantaneous drift close to

$-\alpha Z(t)$. A diffusion process with these parameters is a stationary Ornstein-Uhlenbeck process, with stationary mean zero and stationary variance $\alpha(\beta-\alpha)/\beta^2$. We may conclude that Feller's stochastic model for the logistic process eventually settles down to an equilibrium fluctuation about a constant population size, where this fluctuation is a stationary Gaussian Markov process with variance $N\alpha(\beta-\alpha)/\beta^2$.

This result carries over qualitatively to any finite population stochastic model in which the deterministic model possesses a stable equilibrium.

We turn now to the fluctuating environment type of stochastic model for the logistic process. In this case there are two environmental parameters, the unrestricted growth rate α, and the crowding term β. A stochastic model is obtained by replacing the point (α, β) by a bivariate time series $\{A(t), B(t)\}$. Denoting the population size at time t by $Y(t)$, we have, putting $\alpha(y) = A(t)y - B(t)y^2$ in equation (2.4),

$$\dot{Y}(t) = A(t)Y(t) - B(t)Y^2(t). \tag{8.2}$$

As in the case $\beta = 0$, we proceed by assuming that $A(t) = \alpha + \varepsilon U(t)$ and $B(t) = \beta + \varepsilon V(t)$, where $\varepsilon > 0$ is small. Then equation (8.2) becomes

$$\dot{Y}(t) = \{\alpha + \varepsilon U(t)\}Y(t) - \{\beta + \varepsilon V(t)\}Y^2(t). \tag{8.3}$$

Putting $Y(t) = y(t) + \varepsilon Z(t)$ and using equation (2.5), equation (8.3) yields, dividing through by ε and letting $\varepsilon \to 0$,

$$\dot{Z}(t) = \{\alpha - 2\beta y(t)\}Z(t) + y(t)\{U(t) - y(t)V(t)\}. \tag{8.4}$$

The linear first-order differential equation (8.4) may be solved to yield

$$Z(t) = \int_0^t \exp\left\{-2\beta\int_u^t y(v)dv\right\} y(u)\{U(u) - y(u)V(u)\}\,du. \tag{8.5}$$

When t is large, the right-hand side of equation (8.5) may be approximated by replacing $y(t)$ by its stationary value α/β. We obtain

$$Z(t) \sim \frac{\alpha}{\beta}\int_0^t e^{-2\alpha(t-u)}\left\{U(u) - \frac{\alpha}{\beta}V(u)\right\}du. \tag{8.6}$$

The variance of the right-hand side of (8.6) is, since $E\{U(u)\} = E\{V(u)\} \equiv 0$,

$$v(t) = (\alpha/\beta)^2 \int_0^t du \int_0^t dv\, e^{-2\alpha(t-u)}e^{-2\alpha(t-v)}\gamma(v-u), \tag{8.7}$$

where $\gamma(\cdot)$ is the covariance function of the stationary time series $U(u) - (\alpha/\beta)V(u)$. Transforming variables in the integrand by writing $v = w + u$, (8.7) becomes, integrating with respect to u,

$$v(t) = \frac{\alpha}{2\beta^2}\int_0^t \gamma(w)e^{-2\alpha w}\{1 - e^{-4\alpha(t-w)}\}\,dw. \tag{8.8}$$

It may be noted that $v(t)$ approaches a finite constant as $t \to \infty$. From this we may conclude that the error in the model due to small environmental disturbances is asymptotically a stationary time series. This result is similar to the corresponding result for the finite population effect. As in that case, the conclusion carries over qualitatively to more general models with randomly fluctuating environments.

9. Stochastic models for species interaction

The Volterra and MacArthur models of sections 4 and 5 possess stochastic analogues which are similar to those for the univariate case. If $\boldsymbol{u}(t) = y(t) - \bar{y}$ denotes the departure from the equilibrium solution, for small departures we may write

$$\dot{\boldsymbol{u}}(t) = A\boldsymbol{u}(t). \tag{9.1}$$

Thus, to examine the behaviour of a stochastic species interaction model when the population is near equilibrium, we need only consider the linear model (9.1).

We can obtain a stochastic version of (9.1) by replacing $\boldsymbol{u}(t)$ by a multivariate birth and death process, or by assuming that A is a matrix-valued random process. Multivariate birth and death processes were considered by Kingman (1968). McNeil and Schach (1972) have suggested that stationary vector birth and death processes converge, when suitably normalized and scaled, to multivariate Ornstein-Uhlenbeck processes. Using these results it may be shown that if a multivariate birth and death process is chosen as a model for population growth and interaction, when the population is large it may be approximated by a deterministic part satisfying equation (9.1) plus a relatively small Markovian, Gaussian, stationary vector process.

The other kind of stochastic model, in which A is a matrix-valued random process, has been considered by Sykes (1969). Here the application was the Leslie demographic projection model. Sykes investigated, instead of (9.1), the discrete generation model

$$\boldsymbol{u}_{t+1} = (A + \boldsymbol{\Delta}_t)\boldsymbol{u}_t, \tag{9.2}$$

where $\boldsymbol{\Delta}_t$ is matrix white noise, that is,

$$E\{\boldsymbol{\Delta}_t \boldsymbol{\Delta}_{t+s}\} = 0 \text{ for } s \neq 0. \tag{9.3}$$

As in the univariate case, more general results concerning the covariance structure of $\boldsymbol{u}_t$ are obtainable if the disturbance $\boldsymbol{\Delta}_t$ is presumed small.

10. Conclusion

To conclude this investigation, the following comments may be appropriate. Mathematical models for population growth and species interaction have traditionally been separated into two categories, deterministic models and stochastic models. We have suggested that stochastic models may themselves be classified into one (or both) of two categories, depending on the nature of the stochastic error. In the first category, the error is due to the finiteness of the population; in the second, it is due to unpredictable fluctuations in the environment. In both cases, the deterministic model may be regarded as an extreme case of the stochastic model. In the first case, allowing the population size to become infinite yields the deterministic model as a law of large numbers; in the second case the deterministic model appears, as one would expect, as a limiting case when the environment error approaches zero.

These asymptotic results may be used to obtain quantitative approximations to random effects in ecological models. Since these errors are asymptotically Gaussian, knowledge of their variances enables one to obtain asymptotically valid confidence intervals for population sizes. The results may also be used to obtain the approximate distribution of other quantities of interest such as the time to extinction.

References

Billingsley, P. (1968). *Convergence of Probability Measures*. New York: Wiley.

Feller, W. (1939). Die Grundlagen der Volterraschen Theorie des Kampfes ums Dasein in wahrscheinlichkeitstheoretischer Behandlung. *Acta Biotheoretica* **5,** 11–40.

Galton, F. & Watson, H. W. (1874). On the probability of the extinction of families. *J. Anthrop. Inst.* **4,** 138–44.

Kingman, J. F. C. (1969). Markov population processes. *J. Appl. Prob.* **6,** 1–18.

Leslie, P. H. (1945). On the use of matrices in certain population mathematics. *Biometrika* **33,** 183–212.

Lotka, A. J. (1925). *Principles of Physical Biology*. Baltimore: Williams and Wilkins, reprinted (1959) as *Principles of Mathematical Biology*. New York: Dover.

MacArthur, R. H. (1972). *Geographical Ecology*. New York: Harper and Row.

McKendrick, A. G. (1914). Studies on the theory of continuous probabilities, with special reference to its bearing on natural phenomena of a progressive nature. *Proc. London Math. Soc.* (2) **13,** 401–16.

McNeil, D. R. & Schach, S. (1973). Central limit analogues for Markov population processes (with discussion). *J. Roy. Statist. Soc. B* **35,** 1–23.

Sharpe, F. R. & Lotka, A. J. (1911). A problem in age-distribution. *Phil. Mag.* (6) **21,** 435–8.

Sykes, Z. M. (1969). Some stochastic versions of the matrix model for population dynamics. *J. Am. Statist. Ass.* **325,** 111–30.

Volterra, V. (1926). Variazioni e fluttuazioni del numero d'individui in specie animali conviventi. *Mem. Accad. Lincei* **2,** 31–113 (translated in Chapman, R. N. (1931). *Animal Ecology*. New York: McGraw-Hill.).